MINERALS
HANDBOOK
1982-83

MINERALS
HANDBOOK
1982-83

Phillip Crowson

M

First published by MACMILLAN PUBLISHERS LTD (Journals Division) 1982
Distributed by Globe Book Services Ltd
Canada Road, Byfleet, Surrey KT14 7JL, England

ISBN 978-1-349-06331-4   ISBN 978-1-349-06329-1 (eBook)
DOI 10.1007/978-1-349-06329-1

CONTENTS

INTRODUCTION

This handbook has developed from a study on Non-Fuel Minerals and Foreign
Policy prepared in 1978 for the Royal Institute of International Affairs. A
supplement to the paper provided some "basic data essential to any informed
discussion on the supply of mineral raw materials to the United Kingdom and
the European Community". The study was published in a period when
discussion of mineral procurement policies was emerging from specialist
groups into a broader forum. The statistical supplement apparently met a
need and it was therefore updated and reissued in an expanded form in early
1980. The present handbook is a further extension of that original work.
Its aim is to provide a snapshot of the basic characteristics of the supply
and demand of thirty-seven minerals and metals. Its predecessors
concentrated on the European Community but the tables have here been
extended to cover Japan and the United States. The handbook is not intended
as a substitute for the many excellent statistical publications listed in
the Appendix, from which its data are derived, but merely as an introductory
guide mainly for the non-specialist. It draws together in a convenient form
information that is scattered over a very wide range of primary sources.
The numerous statistical caveats, qualifications and footnotes accompanying
the original sources of the statistics have been omitted. The tables would
otherwise have been swamped in a sea of footnotes.

The earlier versions emphasised that "the real world is invariably far more
complex than simple tables might suggest, and the tendency to latch on to
published statistics as if they were unshakeable truths should be avoided".
To illustrate this point "statistics on reserves of any mineral involve
extensive inference from incomplete data and extensive judgement, not just
about the technical characteristics of ore deposits, but also about their
economics". Even many estimates of production and trade carry wide margins
of error.

The book's layout is straightforward; the introductory summary tables are
followed by separate sections on each of the thirty-seven minerals. The
summary tables mainly bring together data contained in the detailed
sections, but with the addition of a short table (Table 3) showing the
historic growth in reserves of several minerals. A tabulation of the
available data on the western world's investment in minerals is also
included as Table 8. The other tables summarise aspects of mineral
industries that influence public policy. Thus Table 2 shows two measures of
the adequacy of reserves, Table 4 contains estimates of import dependence in
the main areas, and Table 7 shows South Africa's role in the world mineral
industries. South Africa is singled out for a separate summary table solely
because the West's apparent dependence on South Africa dominates much of the
discussion about mineral supplies.

The individual sections on each mineral follow a broadly common format. The
varying units of the sources have been converted into metric equivalents for
all the metals, although imperial and apothecaries' measures are the
conventional measures in many cases. Prices are, however, quoted in their
original units. The main sources of the statistics used are summarised in a
section at the end of the book.

The available data, and the specific characteristics of each mineral explain any variations from the standard pattern, which is as follows.

World Reserves

The statistics are taken mainly from the US Bureau of Mines publications. The figures refer mainly to 1980/81. Separate estimates are shown for most relevant countries which are divided into three broad groups: Developed, Less Developed, and Centrally Planned. In most instances a footnote gives a broader estimate of total resources, which include mineral deposits that are not presently economic. As noted earlier, changes in the basic technical and economic assumptions can dramatically alter estimates of reserves; there is no objective measure. Summary Table 3 demonstrates that estimated reserves have in most instances risen over time at least as fast as production.

World Production

Again there are separate figures for each producer country, subdivided into three broad groupings. The sources are broadly similar to those for reserves. The figures are averages for 1979 and 1980. For some commodities figures are given for the mined product and for its main derivatives. Thus there are separate tables for bauxite, alumina and aluminium.

Secondary Production

Where recycled material is important, and the statistics are available, a separate table shows average supplies in the main areas in 1979/80.

The Adequacy of Reserves

Two estimates are given, based mainly on the earlier tables. The first is the static life of identified reserves, assuming that production continues at the 1979/80 level. In most instances production is growing however, and reserves are also tending to rise as new deposits are discovered, and existing mines extend their knowledge. The second figure, to a certain extent, allows for these changes. It shows the ratio of identified resources to cumulative demand between 1981 and 2000 (based mainly on projections by the US Bureau of Mines). Resources in this context include deposits which may not be economic at today's prices, and with existing technical knowledge. Whilst this dynamic ratio is more meaningful than the static reserve/production ratio, it should, nonetheless, be treated with considerable caution.

Consumption

Average consumption in 1979 and 1980 is given for the main consuming areas, and historic annual average compound growth rates, in all cases during the 1970s, and wherever possible for the 1960s as well. The underlying data have been derived from a wide variety of sources.

End Use Patterns

Data are given for the United States in 1980. Figures for other countries are not as readily available. Although consumption patterns differ in detail for the rest of the world these figures are reasonably representative of the main outlets for each material. In some sections, and particularly where there are important non-metallic as well as metallic uses, separate figures are given for the mineral and for its immediate products.

Value of Contained Metal in Annual Production

Total world production is multiplied by the average prices of 1981 to indicate the product's relative importance as an article of commerce. These values, however approximate, are better guides than relative tonnages alone.

Substitutes and Technical Possibilities

Based largely on the reports of the US Bureau of Mines, these two sections pinpoint how supply and demand may change.

Prices

A description of the pricing methods employed, with a table showing annual average prices between 1976 and 1981. For some commodities only one price is given, whilst in other cases there are several. The prices are taken from various trade publications. Each table is supplemented by a chart which shows the trend in a representative price source 1976. The charts plot index numbers (1981 = 100) of money prices and of prices in 'real' 1981 terms. In order to obtain the latter the money prices are deflated by a relevant wholesale price index. This is the United States' index where prices are given in dollars, as they are in most instances.

Marketing Arrangements

A brief description of the structure of the market, and of any relevant international organisations.

Supply and Demand by Main Market Area

Domestic production, trade and consumption in the United Kingdom, European Community, Japan and United States. Domestic production is divided into the main stages where relevant. The sources of net imports (i.e. imports from third countries) are shown, and also dependence on external supplies. The data are mainly averages for 1979 and 1980. The geographical sources of net imports are given as percentages of the total separately for all four groupings. Shares of world consumption and the historic growth of consumption are also included. For most metals the main additional sources to those used for the earlier tables are the relevant trade statistics.

Acknowledgements

The compiler gratefully acknowledges indebtedness to the statistical publications of the US Bureau of Mines, The World Bureau of Metal Statistics, Metallgesellschaft, and the Institute of Geological Sciences in particular. Many other sources, listed at the end of the report, have also been used. The price data are derived mainly from the Metal Bulletin, Metals Week, Industrial Minerals and the Engineering and Mining Journal. Any mistaken interpretations, errors or omissions, are the compiler's sole responsibility. Particular thanks are due to Nobushige Kondo, Julie Markey and Carolyn Evans for their various contributions to the tables and to Lucy Ereira for producing the typescript.

n/a   Not available
c.   approximately
..   Under 1

TABLE 1

PRODUCTION AND RESERVES BY MAIN GEO-POLITICAL GROUPING

% Share in World	Reserves			Primary Production 1979-80		
	Developed	Less Developed	Centrally Planned	Developed	Less Developed	Centrally Planned
Bauxite	30	67	3	42	47	11
Aluminium	-	-	-	69	10	21
Antimony	21	19	60	34	38	28
Asbestos	63	10	27	42	10	48
Barytes	52	34	14	47	33	20
Cadmium (a)	57	31	12	67	9	24
Chromium	69	30	1	42	22	36
Cobalt	13	62	25	15	73	12
Copper	32	56	12	35	42	23
Fluorspar	55	30	15	32	33	35
Gold	62	13	25	66	11	23
Industrial Diamonds	24	68	8	17	53	30
Iron Ore	35	31	34	39	27	34
Lead	63	18	19	52	20	28
Lithium	28	64	8	79	4	17
Magnesite	7	19	74	33	8	59
Manganese	62	11	27	28	27	45
Mercury	63	14	22	38	17	45
Molybdenum	61	30	9	74	14	12
Nickel	27	54	19	41	31	28
Niobium	5	81	15	17	77	6
Phosphate	15	65	20	42	33	25
Platinum Group	83	..	17	50	..	50
Potash	73	3	24	57	3	40
Rhenium	47	46	7	33	50	17
Selenium	38	51	11	n/a	n/a	n/a
Silicon	n/a	n/a	n/a	65	8	27
Silver	51	24	24	38	38	24
Sulphur	39	34	27	55	9	36
Tantalum	19	75	6	n/a	n/a	n/a
Tin	9	66	25	8	77	15
Titanium: Ilmenite	73	25	2)	83	9	8
Rutile	17	81	2)			
Tungsten	26	9	65	30	23	47
Vanadium	52	2	46	60	1	39
Zinc	62	24	14	54	19	27
Zirconium	65	24	11	85	3	12

(a)   Production at refineries.

TABLE 2

THE 'ADEQUACY' OF RESERVES

	Static Reserve Life (years)	Ratio of Identified Resources to Cumulative Primary Demand 1981-2000
Bauxite	277	12.2
Antimony	68	3
Asbestos	21	1.2
Barytes	26	2.1
Cadmium	36	3
Chromium	346	100 approx.
Cobalt (land only)	105	8.5
Copper	63	8
Fluorspar	23	6.9
Germanium	39	over 3
Gold	27	1.8 (excludes above ground stocks)
Industrial Diamonds	21	under 1 (excluding synthetics)
Iron Ore	186	11
Lead	36	2.4
Lithium	very large	over 20
Magnesium	very large	over 20 (excluding brines & sea)
Manganese (land only)	131	10
Mercury	24	4
Molybdenum	93	8
Nickel (land only)	76	5.5
Niobium	329	38
Phosphate	609	over 10
Platinum Group	180	18
Potash	758	over 170
Rhenium	400	66
Selenium	30 to 40	7.5
Silicon	extremely large	extremely large
Silver	24	2.7
Sulphur	32	3.4
Tantalum	75	9
Tin	42	7
Titanium	88	14
Tungsten	52	5.7
Uranium	56	3.9
Vanadium	444	57
Zinc	26	2
Zirconium	70	4.5

TABLE 3

THE GROWTH OF WORLD RESERVES OF SELECTED PRODUCTS

This table shows how estimates of world reserves of four major base metals increased over a thirty year period relative to the rate of growth of world mine production. Figures for these metals are more readily available than for many others, but in most respects the pattern shown is typical; estimated reserves grew at least as fast as production.

(million tonnes contained metal near the end of the relevant decade)

	Copper	Lead	Zinc	Aluminium (a)
1940s	91	31 to 45	54 to 70	1,605
1950s	124	45 to 54	77 to 86	3,224
1960s	280	86	106	11,600
1970s	543	157	240	22,700
% p.a. growth 1950s-1970s	7.5	5 to 5.75	4.75 to 5.25	9.75
% p.a. growth of mine production 1950s-1970s	3.75	1.75	2.75	7

(a) gross weight of bauxite

TABLE 4

IMPORT DEPENDENCE 1979-80 IN PERCENTAGES
Imports as a percentage of domestic consumption plus exports

	United Kingdom	European Community	Japan	United States
Aluminium (inc. bauxite and alumina	78	45	70	73
Antimony	100 (a)	92 (a)	100 (a)	63
Asbestos	100	84	98	82
Barytes	71	18	36	47
Cadmium (refined)	68	32	-	56
Chromium	100	97	99	c.90
Cobalt (a)	100	100	100	100
Copper	75	80	80	20
Fluorspar	10	29	100	86
Germanium (refined)	100	2	100	under 13
Iron Ore	86	87	89 (b)	25
Lead	53	44	47	13
Lithium	100	100	100	-
Magnesium Metal	75	over 61	9	2
Manganese	100	99.5	99	100
Mercury	100 (a)	91 (a)	28	30
Molybdenum	100	100	99	-
Nickel	100	87	100	88
Niobium	100	100	100	100
Phosphate	100	100	100	1
Platinum Group	100 (a)	100 (a)	100 (a)	93
Potash	64	19	100	69
Rhenium	100	100	100	88
Selenium	100	100	100	63
Silicon	100	44	43	20
Silver	100	58	57	43
Sulphur	92	31	-	16
Tantalum	100	100	100	100
Tin	71	93	99	94
Titanium	100	100	100	80
Tungsten	100 (a)	78 (a)	85 (a)	48
Uranium	100	70	100	..
Vanadium	100	100	100	40
Zinc	81	57	48	49
Zirconium	100	100	100	68

(a) Before allowing for secondary recovery.
(b) Most balance from stocks.

TABLE 5

THE HISTORIC GROWTH OF TOTAL MINE PRODUCTION

% p.a. average compound growth rates 1969 to 1979

Aluminium (bauxite)	5.4
Antimony	0.7
Asbestos	3.4
Barytes	5.7
Cadmium	2.0
Chromium	5.9
Cobalt	3.6
Copper	3.0
Fluorspar	2.3
Germanium	2.9
Gold	-2.0
Industrial Diamonds (including synthetics)	6.8
Iron Ore	2.3
Lead	0.8
Lithium	5.3 approx.
Magnesium (all forms)	0.2
Manganese	1.3
Mercury	-4.0
Molybdenum	3.4
Nickel	3.8
Niobium	4.6
Phosphate	4.9
Platinum Group	6.8
Potash	4.2
Rhenium	11.1
Selenium	2.9
Silicon	5.8
Silver	1.5
Sulphur	3.8
Tantalum	0.3
Tin	1.1
Titanium	1.5
Tungsten	3.3
Uranium	8.0
Vanadium	8.0
Zinc	1.2
Zirconium	3.2 (exc. USA)

TABLE 6

COMPARATIVE GROWTH RATES OF CONSUMPTION IN THE 1970s
% p.a. average compound rates 1969/70 to 1979/80 in most cases

	United Kingdom	European Community	Japan	United States
Aluminium (inc. secondary)	-1.6	4.1	7.3	3.2
Antimony (primary)	-7.5	n/a	-6.9	-3.2
Asbestos	-2.8	1.1	1.9	-3.9
Barytes	6.8	-0.6	0.4	8.9
Cadmium	-0.2	1.4	-3.8	-2.1
Chromium	-4.8	6.5	4.1	0.5
Cobalt	-0.5	0.2	0.7	-
Copper	-1.9	1.6	5.0	0.6
Fluorspar	0.2	-0.4	-0.5	-2.5
Germanium	n/a	n/a	2.1	4.0
Gold (industrial uses)	0.2	-1.0	2.2	-2.6
Industrial Diamonds (inc. synthetics)	n/a	n/a	13.9	7.4
Iron Ore	-2.2	-0.1	7.0	-1.9
Lead	-0.6	0.2	3.4	0.6
Lithium	n/a	n/a	11.7	5.2
Magnesium Metal	-0.5	-2	8.5	2.8
Manganese Ore	-2.7	-0.2	1.5	-6.0
Ferro	-4.7	0.5	2.1	-1.6
Mercury	9.7 (a)	n/a	-11.5	-1.4
Molybdenum	-3.5	2.3	4.4	3.1
Nickel	-0.3	3.4	4.3	1.8
Niobium	-5	5 to 8	12.1 (b)	4.1
Phosphate	0.6	1.7	0.8	4.3
Platinum Group	n/a	n/a	10.1	6.3
Potash	-0.7	1.5	1.0	4.5
Rhenium	n/a	n/a	n/a	7.2
Selenium	2.6	n/a	1.0	-5
Silicon	-2	n/a	5.8	2.4
Silver (industrial uses)	-0.7	-1.9	3.9	0.8
Sulphur	-0.7	0.7	-1.4	3.4
Tantalum	n/a	n/a	13.6 (c)	1.8
Tin	-4.9	-1.7	1.9	-1.6
Titanium	-1.3	-	4.3	1.5
Tungsten	-8.3	-6	-4.9	1.6
Uranium (civil usage)	n/a	14.3	25.3	10.6
Vanadium	-3	n/a	8.7	-0.3
Zinc	-3	0.8	2.2	-2.4
Zirconium	-1.7	4.2	8.4	0.2

(a) Primary only.
(b) Ferro niobium only.
(c) Powder only.

TABLE 7

SOUTH AFRICAN SHARES OF WORLD RESERVES AND PRODUCTION
(percentages)

	Reserves	Primary Production 1979-80 Averages
Antimony	7	19
Asbestos	21	5 (but 100% of amosite and crocidolite)
Cadmium	8	(a)
Chromium	68	35
Cobalt	n/a	..
Copper	1	3
Fluorspar	30	10
Gold	51	57
Industrial Diamonds	8	17
Iron Ore	1	3
Lead	4	1
Magnesite	n/a	1
Manganese	53	21
Nickel	3	4
Phosphate	1	2
Platinum Group	81	46
Silicon	n/a	4 (metal and ferro)
Silver	..	1
Sulphur	n/a	1
Tin	1½	1
Titanium:		
Ilmenite	15)	10
Rutile	4)	
Uranium	15 (b)	13 (b)
Vanadium	49	35
Zinc	7	1
Zirconium	22	11

(a) No details available on mine output by country.
(b) Western world only.

TABLE 8

WESTERN WORLD INVESTMENT IN MINERALS

Figures have been converted to US dollars and then deflated to 1980 terms
with an index of US mine equipment costs. Five year totals are given
because of the lumpy nature of investments and erratic annual movements.

US$ million in real 1980 terms Five year totals

	1966-70	1971-75	1976-80
Australia	7745	10295	6200
Canada	8415	10195	7685
South Africa	3575	5530	7530
Europe (a)	1680	2650	1925
United States (inc. coal)	19365	28035	34240
Total Developed Countries	40780	56705	57580
Less Developed Countries (b)	7580	10615	9010
Identified Western world Total	48360	67320	66590

Note:-   The overall total is probably 10 to 20% higher than the figures
shown. They cover investment in all forms of mining (excluding
petroleum and natural gas).

(a)   Major EC mining companies plus total investment in Finland, Greece
and Sweden. Thus Norway, Portugal, Spain and Yugoslavia not
included.

(b)   Investment by Canadian, Japanese, US and EC mining companies in
less developed countries plus Brazil, Chile, Philippines, Zaire and
Zambia. Several major mining countries (e.g. India and Mexico)
excluded.

ALUMINIUM/BAUXITE/ALUMINA

WORLD RESERVES OF BAUXITE
(million dry tonnes and % of total)

Developed			Less Developed			Centrally Planned			Total
Australia	6400	(26.0)	Brazil	2500	(10.2)	China	150	(0.6)	
			Cameroon	1000	(4.1)	Hungary	300	(1.2)	
Greece	700	(2.8)	Ghana	570	(2.3)	USSR	300	(1.2)	
			Guinea	6500	(26.5)				
Yugoslavia	400	(1.6)	Guyana	700	(2.8)				
Other(inc.	130	(0.5)	India	1000	(4.1)				
France & USA)			Indonesia	700	(2.8)				
			Jamaica	2000	(8.1)				
			Sierra Leone	130	(0.5)				
			Surinam	490	(2.0)				
			Others	600	(2.4)				
Totals	7630	(31.1)		16190	(65.9)		750	(3.0)	24570

Total world resources (reserves plus sub-economic and undiscovered deposits) are estimated at 40 to 50,000 million tonnes.

WORLD RESERVES OF BAUXITE IN TERMS OF RECOVERABLE ALUMINIUM CONTENT
(million tonnes of metal and % of total)

Developed			Less Developed			Centrally Planned			Total
Australia	1290	(25.3)	Brazil	555	(10.9)	China	30	(0.6)	
Greece	145	(2.8)	Cameroon	180	(3.5)	Hungary	65	(1.3)	
Yugoslavia	75	(1.5)	Ghana	120	(2.4)	USSR	55	(1.1)	
Other	25	(0.5)	Guinea	1360	(26.7)				
			Guyana	170	(3.3)				
			India	205	(4.0)				
			Indonesia	135	(2.6)				
			Jamaica	415	(8.1)				
			Sierra Leone	25	(0.5)				
			Surinam	120	(2.4)				
			Others	130	(2.5)				
Totals	1535	(30.1)		3415	(67.0)		150	(2.9)	5100

Total world resources (reserves plus sub-economic and undiscovered deposits) are calculated at 8,000 million tonnes of recoverable aluminium, on the basis of present recovery techniques. (The USSR also produces aluminium from alunite and nepheline syenite so that the table under-estimates the USSR's available deposits of aluminium containing minerals).

BAUXITE: WORLD MINE PRODUCTION
('000 tonnes and % of total 1979/80 Averages)

Developed			Less Developed			Centrally Planned			Total
Australia	27584	(31.1)	Brazil	3179	(3.6)	China	1500	(1.7)	
France	1817	(2.0)	D. Republic	565	(0.6)	Hungary	2998	(3.4)	
Greece	2933	(3.3)	Ghana	220	(0.2)	Romania	708	(0.8)	
USA	1690	(1.9)	Guinea	13740	(15.5)	USSR	4600	(5.2)	
Yugoslavia	3075	(3.5)	Guyana	2330	(2.6)				
Others	393	(0.4)	Haiti	518	(0.6)				
			India	1837	(2.1)				
			Indonesia	1138	(1.3)				
			Jamaica	11883	(13.4)				
			Malaysia	654	(0.7)				
			Sierra Leone	587	(0.7)				
			Surinam	4853	(5.5)				
			Others	18	(..)				
Totals	37492	(42.2)		41522	(46.8)		9806	(11.0)	88820

Note: The USSR's production of nepheline syenite and alunite were equivalent to roughly 1.6 million tonnes of bauxite.

ALUMINA: WORLD REFINERY PRODUCTION
('000 tonnes and % of total 1979/80 Averages)

Developed			Less Developed			Centrally Planned			Total
Australia	7331	(22.9)	Brazil	495	(1.5)	China	750	(2.3)	
Canada	981	(3.1)	Guinea	684	(2.1)	Czecho-slovakia	100	(0.3)	
France	1124	(3.5)	Guyana	187	(0.6)	E Germany	41	(0.1)	
W Germany	1376	(4.3)	India	497	(1.6)	Hungary	794	(2.5)	
Greece	493	(1.5)	Indonesia	2276	(7.1)	Romania	500	(1.6)	
Italy	877	(2.7)	Surinam	1321	(4.1)	USSR	2650	(8.3)	
Japan	1748	(5.5)	Taiwan	62	(0.2)				
Spain	29	(0.1)							
Turkey	140	(0.4)							
UK	89	(0.3)							
USA	6630	(20.7)							
Yugoslavia	853	(2.7)							
Totals	21671	(67.7)		5522	(17.2)		4835	(15.1)	32028

PRIMARY ALUMINIUM PRODUCTION
('000 tonnes and % of total 1979/80 Averages)

Developed			Less Developed			Centrally Planned			Total
Australia	286.6	(1.8)	Argentina	131.5	(0.8)	China	355	(2.3)	
Austria	93.6	(0.6)	Bahrain	126.2	(0.8)	Czecho-slovakia	38	(0.2)	
Canada	969.1	(6.2)	Brazil	249.5	(1.6)	E Germany	60	(0.4)	
France	413.5	(2.6)	Cameroon	43.8	(0.3)	Hungary	72	(0.5)	
W Germany	736.3	(4.7)	Dubai (a)	35	(0.2)	N Korea	9	(0.1)	
Greece	143.2	(0.9)	Egypt	110.6	(0.7)	Poland	96	(0.6)	
Ireland	73.5	(0.5	Ghana	178	(1.1)	Romania	229	(1.5)	
Italy	270.2	(1.7)	India	193	(1.2)	USSR	2385	(15.2)	
Japan	1051.0	(6.7)	Iran	13.5	(0.1)				
Netherlands	256.9	(1.6)	S Korea	17.5	(0.1)				
New Zealand	155.2	(1.0)	Mexico	42.9	(0.3)				
Norway	667.6	(4.3)	Surinam	57.5	(0.4)				
South Africa	86.4	(0.6)	Taiwan	59.3	(0.4)				
Spain	323.0	(2.1)	Venezuela	277.7	(1.8)				
Sweden	81.8	(0.5)							
Switzerland	84.7	(0.5)							
Turkey	32.8	(0.2)							
UK	367.0	(2.3)							
USA	4605.2	(29.4)							
Yugoslavia	166.4	(1.1)							
Totals	**10864**	**(69.4)**		**1536**	**(9.8)**		**3244**	**(20.7)**	**15644**

(a) 1980 only

ALUMINIUM RECOVERED FROM SCRAP : WESTERN COUNTRIES
('000 tonnes 1979/80 Averages)

European Community	1097
Japan	768
United States	1576
Other Countries	391
Total	3832

RESERVE/PRODUCTION RATIOS FOR BAUXITE

Static Reserve life (years)	277
Ratio of resources to cumulative demand 1981-2000 (includes identified sub-economic, hypothetical and speculative resources)	12.2 : 1

CONSUMPTION OF PRIMARY ALUMINIUM

	'000 tonnes 1979/80 Averages	Growth rate % p.a.	
		1960-70	1970-80
European Community (Ten)	2959	7.5	4.2
Japan	1720	20.7	7.1
USA	4745	7.8	2.8
Others	2876	4.7	7.5
Total Western World	12300	9.3	4.6
Total World	15642	9.2	4.7

END USE PATTERNS 1980 (USA) %

Bauxite/Alumina

Aluminium Metal: 90%
Refractories, chemicals, abrasives and other products: 10%

Aluminium

Packaging	30
Building	21
Transport	18
Electrical	10
Machinery & Equipment	8
Others	13

VALUE OF CONTAINED METAL IN ANNUAL PRODUCTION

$27.4 billion (primary metal) at 1981 world producer price
$19.7 billion (primary metal) at 1981 average LME price

SUBSTITUTES

Bauxite/Alumina

Replacement of alumina abrasives with synthetic diamonds, silicon carbide and aluminium-zirconium oxide. Clays and alumina can substitute for bauxite for making aluminous refractories and chemicals.

Aluminium

Plastics and steel compete for many applications, notably for machinery, household appliances, and, with glass and paper, for the container market. Magnesium, titanium and composites compete in the transport and structural industry whilst wood is becoming increasingly important in the construction industry. Copper can be used in many applications.

Potential for substitutes often limited by relative weight (steel) or cost (titanium, magnesium).

TECHNICAL POSSIBILITIES

Bauxite/Alumina

Development continuing on alternative raw materials including coal wastes, anorthosite, clay and shale. Political considerations likely to be more important.

Possible development of other refractories, using nitrides and borides of titanium and zirconium.

Chemical use may be limited by development of chemicals or processes for recycling water.

Aluminium

Energy costs a significant constraint on industrial country production. Advances in alumina reduction methods should help keep aluminium competitive.

PRICES

Bauxite

Prices historically mainly transfer prices within integrated alumina or aluminium producers or negotiated under long term contracts and market prices unrepresentative. In 1974 Caribbean governments substantially boosted effective prices by tax arrangements linked to final aluminium price. Levy has caused some curtailment of bauxite operations in this area.

Aluminium

	1976	1977	1978	1979	1980	1981
cents/lb						
World Producer (Alcan Price)	43.9	50.3	52.9	63.5	77.8	79.5
US Dealer-range	34-45	44-51	48-56.5	57-78	67-93.5	47-68
LME Cash	-	-	-	72.5	81	57
£/tonne						
LME Cash Monthly Average Range	-	-	-	630.8-870.0	610.8-930.3	568.8-676.6

Combination of producer prices and free dealer market. Latter fluctuates widely. Producer discounting from list price widespread in times of low demand, though production cutbacks come into force also. Cost pressures of bauxite taxes and energy costs intense. LME quotation introduced in 1979. World price diverges from US producer price when price controls in force in USA (eg: 1979).

MARKETING ARRANGEMENTS

Bauxite/Alumina

Sales mainly between integrated aluminium producers. Trend towards nationalisation of producing companies in developing countries, especially in Caribbean. International Bauxite Association formed in 1974 - Jamaica, Guyana, Haiti, Indonesia, Surinam, Dominican Republic, Ghana, Guinea, Sierra Leone, Yugoslavia, and Australia are members. Pools price and market information with objective of 'fair and reasonable returns'. Search for minimum price arrangements so far unsuccessful. Brazil to join IBA in near future and likely to join Australia in opposition to strict pricing policies and cartelisation. USA currently adding bauxite to its stockpile.

Aluminium

Market previously dominated by Alcan, Alcoa, Kaiser, Reynolds, Alusuisse, Pechiney and their associate/subsidiaries. Substantial degree of vertical integration from mine to fabricated products. Growth of free market with rise of independent producers, especially in Third World energy rich countries. Location of new smelters likely to be in these countries or Australia. These various changes, compounded by recession are breaking down the oligopolistic structure of the industry.

Index Numbers 1981 = 100

The solid line gives prices in money terms and the dotted line gives prices in 'real' 1981 terms

ALUMINIUM
World Producer Price

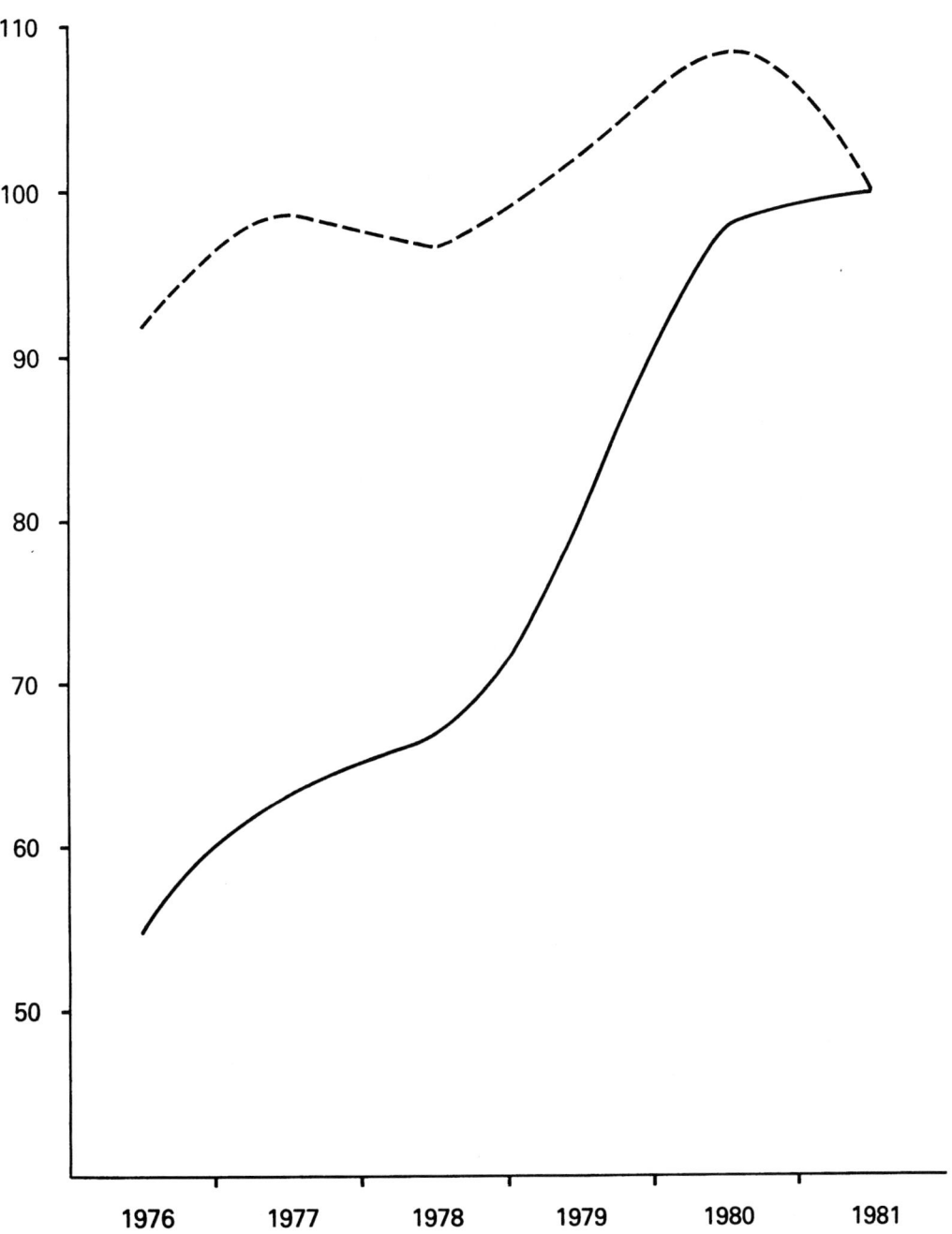

SUPPLY & DEMAND FOR BAUXITE BY MAIN MARKET AREA

	UK	EC (Ten)	Japan	USA
Production (1979/80 Averages) ('000 tonnes)	nil	5031	nil	1690
Net Imports (1979/80 Averages) ('000 tonnes)	276	7705	5152	14251
Source of Net Imports (%)				
Australia	4	45	67	-
European Community	17	-		
Brazil		-		3
China		2		
Dominican Republic				4
Ghana	67	3		
Guinea	1	39		29
Guyana		2		4
Haiti		-		4
Indonesia		1	21	-
Jamaica				45
Malaysia		..	9	
Sierra Leone	6	6		1
Surinam	3	1		10
Others		1	3	
Net Exports (1979/80 Averages)	1	1507	nil	18
Consumption (1979/80 Averages) ('000 tonnes)	275 (apparent)	11229 (apparent)	4679	15829
Import Dependence				
Imports as % of consumption	100	69	100	90
Imports as % of consumption and net exports	100	60	100	90
Share of World Consumption %				
Total World	..	12	5	18
Consumption Growth % p.a.				
1970s	-1.1	6	6.6	-

SUPPLY AND DEMAND FOR ALUMINA BY MAIN MARKET AREA

	UK	EC (Ten)	Japan	USA
Production (1979/80 Averages) ('000 tonnes)	89	3959	1741	6630
Net Imports (1979/80 Averages)	686	1366	734	4098
Source of Net Imports (%)				
Australia		15	98	77
Canada				1
European Community	6		2	
USA		..		
Yugoslavia		2		
Guinea		7		
Guyana	8	4		
Jamaica	68	34		15
Surinam	16	32		6
Others	1	6		1
Net Exports (1979/80 Averages) ('000 tonnes)	8	490	286	994
Consumption (1979/80 Averages) ('000 tonnes)	767 (apparent)	4835 (apparent)	2009	9734 (apparent)
Import Dependence				
Imports as % of consumption	89	28	37	42
Imports as % of consumption and net exports	89	26	32	38
Share of World Consumption %				
Total World	2	14	6	29
Consumption Growth % p.a.				
1970s	-1.1	6	6.4	-

SUPPLY AND DEMAND FOR ALUMINIUM METAL BY MAIN MARKET AREA

	UK	EC (Ten)	Japan	USA
Production (1979/80 Averages) ('000 tonnes)				
Primary Metal	367	2187	1051	4605
Secondary Metal	155	953	694	1294
Total	522	3140	1745	5899
Net Imports (1979/80 Averages) ('000 tonnes)	173	898	815	517
Source of Net Imports (%) (a)				
European Community	13	–	1	2
USA	2	7	24	–
Iceland	12	6	–	–
Norway	63	48	2	4
Spain	(	5	3	)
Switzerland	(2	5	..	) 1
Other Europe	(	3	1	)
Ghana	..	3	4	17
Other Africa	..	4	2	..
Argentina	..	2	2	–
Canada	2	4	12	74
Surinam	2	4	1	1
Venezuela	3	..	15	1
Bahrain	–	..	4	–
Australia	–	–	4	–
New Zealand	–	–	12	–
Eastern Bloc	..	9	12	–
Other Countries	1	..	..	–

(a) Note that percentages have been rounded and may not total 100

	UK	EC (Ten)	Japan	USA
Net Exports (1979/80 Averages) ('000 tonnes)	199	122	8	406
Consumption (1979/80 Averages) ('000 tonnes)				
Primary Metal	413	2959	1719	4741
Secondary Metal	94	923	615	1294
Total	507	3882	2334	6035

	UK	EC (Ten)	Japan	USA
Import Dependence				
Imports as % of consumption	34	23	35	9
Imports as % of consumption and net exports	25	22	35	8
Share of World Consumption % (Primary and Secondary Metal)				
Western World	3.2	24.5	14.7	38.1
Total World	2.5	19.4	11.7	30.2
Consumption Growth % p.a.				
Primary Metal 1970s	1.6	4.2	7.1	2.8
Total Metal 1970s	-1.6	4.1	7.3	3.2

****

ANTIMONY

WORLD RESERVES
('000 tonnes and % of total)

Developed			Less Developed			Centrally Planned			Total
Australia	135	(3)	Bolivia	370	(8)	China	2370	(53)	
Canada	65	(1)	Malaysia	120	(3)	Czecho-slovakia	45	(1)	
Italy	120	(3)	Mexico	225	(5)	USSR	270	(6)	
S Africa	325	(7)	Peru	65	(1)				
Turkey	110	(2)	Thailand	90	(2)				
USA	120	(3)							
Yugoslavia	90	(2)							
Totals	965	(21)		870	(19)		2685	(60)	4520

Identified world resources estimated at 5.2 million tonnes.

WORLD MINE PRODUCTION
('000 tonnes metal and % of total 1979/80 Averages)

Developed			Less Developed			Centrally Planned			Total
Australia	1.56	(2.4)	Bolivia	14.24	(21.6)	China	9.98	(15.1)	
Austria	0.56	(0.8)	Brazil	0.20	(0.3)	Czecho-slovakia	0.30	(0.5)	
Canada	2.55	(3.9)	Burma	0.73	(1.1)	USSR	8.16	(12.3)	
Italy	0.83	(1.3)	Guatemala	0.70	(1.1)				
S Africa	12.36	(18.7)	Honduras	0.15	(0.2)				
Spain	0.50	(0.8)	Malaysia	0.38	(0.6)				
Turkey	1.85	(2.8)	Mexico	2.89	(4.4)				
USA	0.48	(0.7)	Morocco	1.98	(3.0)				
Yugoslavia	1.79	(2.7)	Pakistan	..					
			Peru	0.77	(1.2)				
			Thailand	2.89	(4.4)				
			Zimbabwe	0.24	(0.4)				
Totals	22.48	(34.0)		25.17	(38.1)		18.44	(27.9)	
66.09									

SECONDARY PRODUCTION

Sizeable tonnages of antimony are contained in recycled antimonial lead, on which the available statistics are incomplete. Total secondary recovery averaged 19,980 tonnes of contained antimony in the United States in the two years 1979, and 1980, and 4,602 tonnes in the United Kingdom. Changes in battery technology are reducing this source of supply.

RESERVE/PRODUCTION RATIOS

Static Reserve life (years)   68
Ratio of identified resources
to cumulative demand 1981-2000   3 : 1

CONSUMPTION

	1979/80 Averages tonnes	% p.a. growth 1970s
European Community (Ten)	8600 (a)	falling fast
United States	10429 (b)	-3.2
Japan	1453.5 (metal)	-6.9

(a) Estimated Sb content of primary
(b) Reported primary. Apparent total, including secondary was 34,774 tonnes Sb content with a 1970s' growth rate of 0.2% p.a.

END USE PATTERNS 1980 (USA) %

Transport   15
Flame retardants   55
Chemicals   5
Ceramics & Glass   15
Other   10

VALUE OF CONTAINED METAL IN ANNUAL PRODUCTION

$188 million (at average 1981 European Free Market metal price)

SUBSTITUTES

Tin, calcium, tellurium, cadmium substitute as hardeners for lead in batteries.

Antimony can be replaced by organic compounds or oxides of iron and titanium in flame retardants and by tellurium and selenium in rubber manufacture.

Plastics or stainless steel products can replace enamel coated products. Titanium, zinc, chromium, tin and zirconium may be substituted in paints, pigments and enamels.

TECHNICAL POSSIBILITIES

More specialised plastics.

Advances in storage battery construction bringing displacement of antimony.
Development of electric vehicles could utilise high-antimony batteries for
deep-cycling characteristics.

PRICES

	1976	1977	1978	1979	1980	1981
Ore. Lump sulphide ore 60% Sb cif $/metric ton unit Sb range	21- 27.25	16- 26.5	15.75- 20	18.75- 24.75	23- 25	20- 25
Metal. European Free Market Regulus 99.6% $/tonne	3251.1	2529.2	2382.1	3062.4	3295.1	2838.6

Supply/demand balance important and brings fluctuating prices. Free market
most important though some producer pricing for antimonial lead.

MARKETING ARRANGEMENTS

Mixture of state-owned production (Comibol in Bolivia, Russia and China) and
large companies (eg: Consolidated Murchison in S Africa). Bolivia, Peru,
Thailand and Turkey formed International Producers Group in 1981 and hope to
stabilise prices through production control. Little has been achieved so
far however. Disposal of 2700 t. of antimony from USA stockpile overhangs
market at the moment.

Index Numbers 1981 = 100

The solid line gives prices in money terms and the dotted line gives prices in 'real' 1981 terms

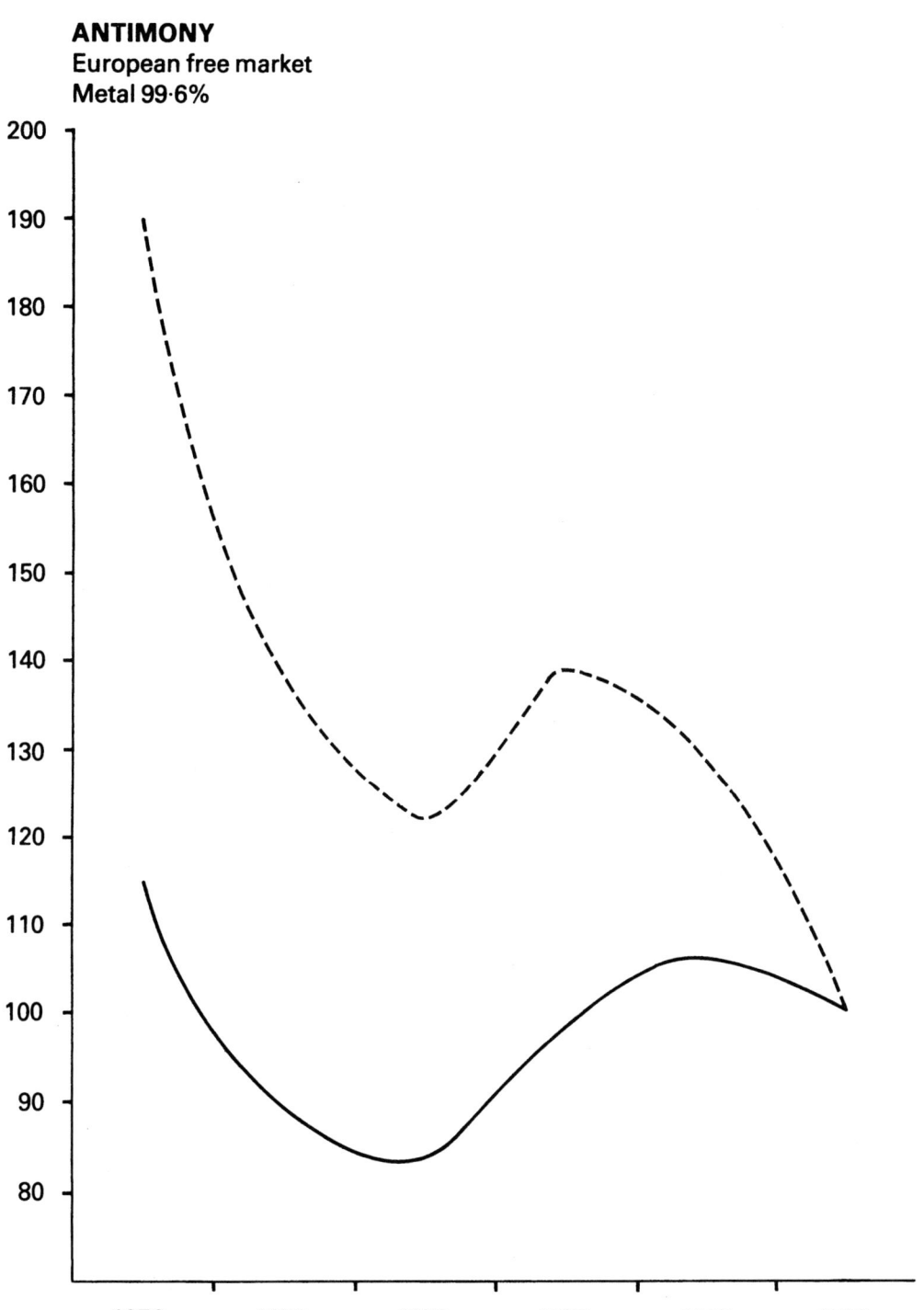

ANTIMONY
European free market
Metal 99·6%

SUPPLY AND DEMAND BY MAIN MARKET AREA

	UK	EC (Nine)	Japan	USA
Production (1979/80 Averages) (tonnes)				
Mine	-	831	nil	483
Metal (primary)		exceeds		
	1000 (est)	2875	433.5	1428 (Sb)
Oxides	n/a	n/a	n/a	12520 (Sb)
Net Imports (1979/80 Averages) (tonnes)				
Ores and concentrates	n/a	18555	6849	12151
Unwrought metal	350	1216 (excl. UK)	2382	2546
Oxide	391	710	-	11749
Total Sb content	680	10595	5629	22141
Wrought metal	n/a	249	-	

Source of Net Imports (%)
Ores and Concentrates

	UK	EC (Nine)	Japan	USA
Australia		8		
Austria		1		
Canada	n/a	5		16
European Community		-	-	2
S Africa		6		7
Spain			1	-
Bolivia		29	6	23
Chile		3	-	6
China		7	91	
Guatemala		5		
Malaysia		2		
Mexico				39
Morocco		3		
Peru		3		
Thailand		21		4
Turkey		3		
Uruguay				2
Vietnam			1	
Others		4		1

	UK	EC (Nine)	Japan	USA
Unwrought Metal				
European Community	41			9
S Africa			7	
Spain		16		
Bolivia			85	41
Chile				2
China	44	75		32
Mexico				10
Thailand	-	-	5	-
Others	15	9		6
Oxide				
European Community	99			20
Bolivia		71		7
China				16
S Africa				55
Others	1			2

Net Exports (1979/80 Averages)
(tonnes)

	UK	EC (Nine)	Japan	USA
Ores and concentrates	610	599	-	-
Unwrought metal	10	60	38	425
Oxides	3000	6301	-	630
Wrought metal	220	201	10	-

Consumption (1979/80 Averages)
(tonnes)

	UK	EC (Nine)	Japan	USA
	600 (primary metal) c.1800 (all primary Sb content) c.6500 (total Sb inc. secondary)	8600 (est. primary Sb content)	1453.5 (metal)	10429 (reported primary Sb content) 34774 (total apparent inc. secondary Sb content)

	UK	EC (Nine)	Japan	USA
Import Dependence				
Imports as % of consumption	100 (primary)	90 (primary)	100 (primary)	64 (inc. secondary)
	28 (all forms)			
Imports as % of consumption and net exports	100 (primary)	92 (primary)	100 (primary)	63 (inc. secondary)
Share of World Consumption %				
Total World	3	13	2(+)	16(+)
Consumption Growth % p.a.				
1970s	-7.5 (primary)	n/a but large fall in 1970s	-6.9 (metal)	-3.2 (reported primary) 0.2 (apparent total)

****

ASBESTOS

WORLD RESERVES
(million tonnes and % of total)

Developed			Less Developed			Centrally Planned			Total
Canada	37	(35.6)	Total	10	(9.6)	Total	28	(26.9)	
S Africa	22	(21.2)							
USA	4	(3.8)							
Others	3	(2.9)							
(Australia,									
Cyprus, Japan,									
Yugoslavia)									
Totals	66	(63.5)		10	(9.6)		28	(26.9)	104

The world's identified resources total 200 million tonnes, and hypothetical resources include an additional 45 million tonnes.

WORLD MINE PRODUCTION
('000 tonnes and % of total 1979/80 Averages)

Developed			Less Developed			Centrally Planned			Total
Australia	80	(1.6)	Brazil	139	(2.9)	Bulgaria	1	(..)	
Canada	1392	(28.7)	India	38	(0.8)	China	250	(5.2)	
Cyprus	35	(0.7)	Korea	14	(0.3)	USSR	2085	(43.0)	
Italy	144	(3.0)	Swaziland	35	(0.7)				
Japan	3	(0.1)	Zimbabwe	256	(5.3)				
S Africa	260	(5.4)	Others	5	(0.1)				
Turkey	19	(0.4)							
USA	87	(1.8)							
Yugoslavia	11	(0.2)							
Totals	2031	(41.8)		487	(10.0)		2336	(48.2)	4854

South Africa is the predominant source of amosite and crocidolite.

RESERVE/PRODUCTION RATIOS

Static Reserve life (years)   21
Ratio of identified resources
to cumulative demand 1981-2000   1.2 : 1

CONSUMPTION

	1979/80 Averages '000 tonnes	% p.a. Growth rates 1970s
European Community (Ten)	869 (apparent)	1.1
Japan	278.5	1.9
United States	461	-3.9

END USE PATTERNS 1980 (USA) %

Asbestos-cement pipe	40
Flooring products	25
Friction products	12
Roofing products	7
Packing and Gaskets	3
Others	13

VALUE OF ANNUAL PRODUCTION

$3 billion approx. (based on 1981 average prices)

SUBSTITUTES

Environmental and health problems associated with asbestos have led to search for substitutes but few can give both chemical and physical characteristics at the same cost. Synthetic inorganic fibres are becoming commercially available but few compete seriously. Glass-reinforced cement is a possibility but has a number of drawbacks.

Substitution is possible in many end uses particularly asbestos-cement products where ceramic and new plastic materials available.

TECHNICAL POSSIBILITIES

Changes in manufacturing methods to reduce health hazards.

Potential for use in high strength asphalt paving materials. Use as a reinforcing agent for lightweight plastics will grow as energy conservation increases in car and steel-consuming industries.

PRICES

	1976	1977	1978	1979	1980	1981
Canadian Chrysotile Fibre C$/short ton (range)						
Group 3 (Spinning fibre)	891-1668	1016-1668	1016-1668	1016-1668	1016-1770	1051-1979
Group 4 (Shingle fibre)	492-945	561-945	561-1011	561-1011	687-1135	765-1247
Group 7 (Refuse/Shorts)	89-195	101-195	101-218	101-218	113-240	120-252

Producer pricing on fixed contracts with discounting. Price depends on grade.

MARKETING ARRANGEMENTS

Most of Western World production of chrysotile controlled by few large firms, with considerable vertical integration from mine to manufactured products.

Index Numbers 1981 = 100

The solid line gives prices in money terms and the dotted line gives prices in 'real' 1981 terms

ASBESTOS
Canadian Chrysotile Fibre, Grp.7
(Lower limit)

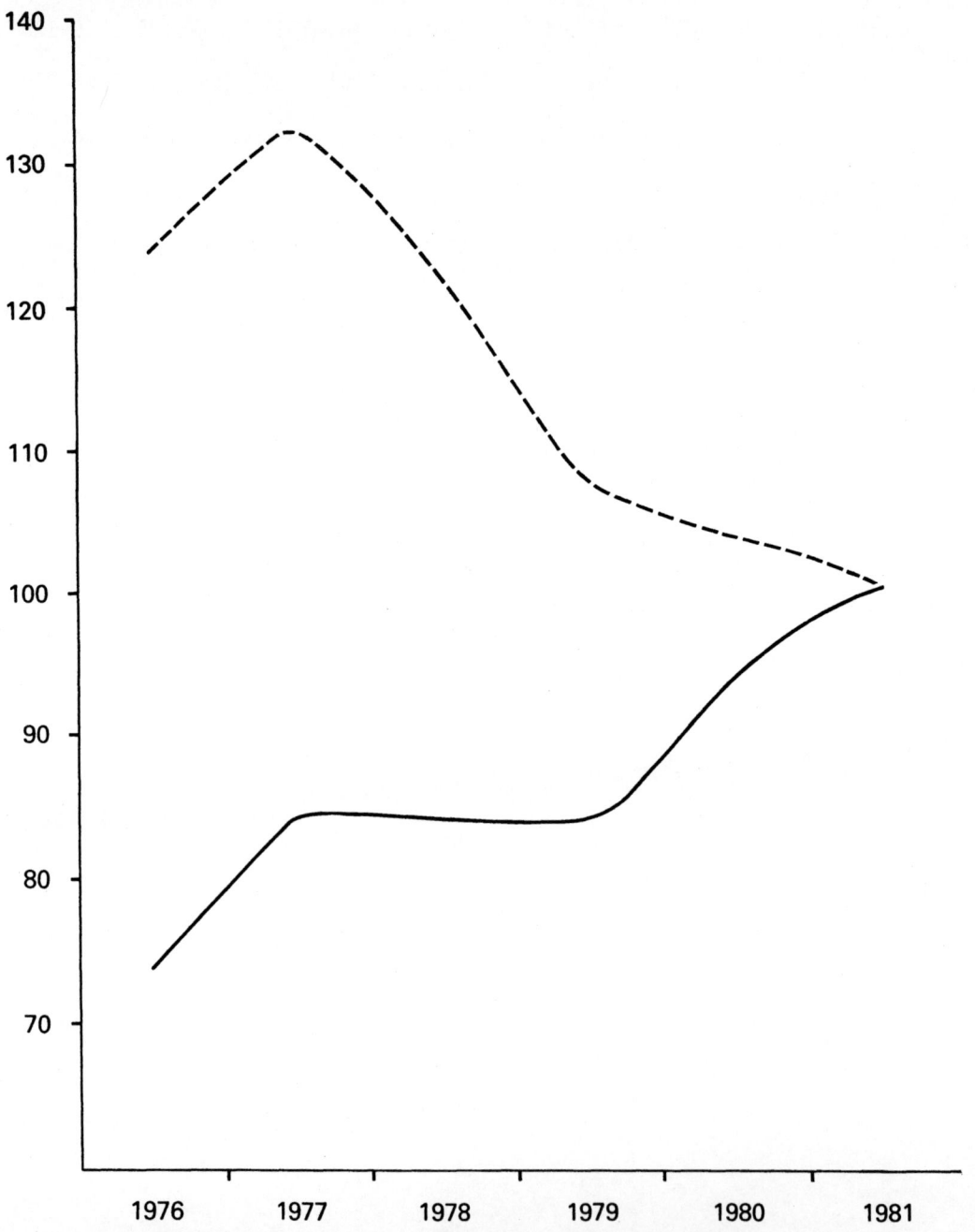

SUPPLY AND DEMAND BY MAIN MARKET AREA

	UK	EC (Ten)	Japan	USA
Production (1979/80 Averages) '000 tonnes	-	144 (Italy)	6	86.5
Net Imports (1979/80 Averages) '000 tonnes	105.7	757.7	298.5 of which) crocidolite) and)11 amosite)	420
Source of Net Imports (%)				
Australia	1	..		
Canada	80	72	40	97
Cyprus	7	2		
European Community	2			
S Africa	9	14	30	3 (a)
USA		..		
USSR		9	14	
Zimbabwe	1	..		
Others		2	16	

(a) 97% of amosite and crocidolite

	UK	EC (Ten)	Japan	USA
Net Exports (1979/80 Averages) ('000 tonnes)	0.9	32.4	nil	48.5
Consumption (1979/80 Averages) ('000 tonnes)	105 (apparent)	869 (apparent)	278.5	461
Import Dependence				
Imports as % of consumption	100	87	98	91
Imports as % of consumption and net exports	100	84	98	82
Share of World Consumption %				
Total World (approx.)	2	18	6	9
Consumption Growth % p.a.				
1970s	-2.8	1.1	1.9	-3.9

****

BARYTES

WORLD RESERVES
('000 tonnes and % of total)

Developed			Less Developed			Centrally Planned			Total
Canada	14500	(8)	Brazil	3500	(2)	China	9000	(4)	
France	5400	(3)	India	27200	(14)	USSR	9100	(5)	
Germany	5400	(3)	Mexico	9100	(5)	Others	9100	(5)	
Greece	1800	(1)	Morocco	2700	(1)				
Ireland	8200	(4)	Peru	6300	(3)				
Italy	5400	(3)	Thailand	9100	(5)				
Turkey	4500	(2)	Others	8400	(4)				
USA	50000	(26)							
Yugoslavia	2700	(1)							
Others	1800	(1)							
Totals	99700	(52)		66300	(34)		27200	(14)	193200

Total world resources are believed to be roughly 1800 million tonnes of which only 350 million are identified.

WORLD MINE PRODUCTION
('000 tonnes and % of total 1979/80 Averages)

Developed			Less Developed			Centrally Planned			Total
Canada	77	(1.0)	Algeria	90	(1.2)	China	590	(8.0)	
France	231	(3.1)	Argentina	57	(0.8)	Czecho-slovakia	64	(0.9)	
W Germany	156	(2.1)	Brazil	108	(1.5)	E Germany	18	(0.2)	
Greece	48	(0.7)	Chile	215	(2.9)	N Korea	109	(1.5)	
Ireland	330	(4.5)	India	367	(5.0)	Poland	93	(1.3)	
Italy	216	(2.9)	Iran	166	(2.3)	Romania	88	(1.2)	
Japan	53	(0.7)	Mexico	241	(3.3)	USSR	499	(6.8)	
Spain	73	(1.0)	Morocco	302	(4.1)				
Turkey	129	(1.8)	Peru	445	(6.1)				
UK	48	(0.7)	Thailand	342	(4.7)				
USA	1976	(26.9)	Others	126	(1.7)				
Yugoslavia	44	(0.6)							
Others	39	(0.5)							
Totals	3420	(46.6)		2459	(33.5)		1461	(19.9)	7340

RESERVE PRODUCTION RATIOS

Static Reserve life (years)   26
Ratio of identified resources
to cumulative demand 1981-2000   2.1 : 1

CONSUMPTION

	1979/80 Averages '000 tonnes	% p.a. Growth rates 1970s
European Community (Ten)	837	-0.6
Japan	90	0.4
United States	3117	8.9

END USE PATTERNS 1981 (USA) %

Drilling   95
Chemicals, Glass, Paint,
Rubber   5

VALUE OF ANNUAL PRODUCTION

$0.4 billion (at 1981 average prices)

SUBSTITUTES

Drilling mud substitutes include celestite, iron ores, synthetic hematite and ilmenite.

TECHNICAL POSSIBILITIES

Reclaiming and recycling of drilling muds would decrease requirement for new supplies.

New energy sources and development of enhanced oil recovery techniques would reduce need for conventional oil and gas and hence bring drop in drilling activity.

PRICES

	1976	1977	1978	1979	1980	1981
Drilling mud grade, 4.2 SG CIF £/tonne (range)	26-31	33-41	39-41	39-45	40-44	42-50
Ground white paint grade 96-98% $BaSO_4$ £/tonne (range)	69-79	69-80	70-80	70-105	85-105	85-105

Usually long term supply contracts. Transport costs important.

MARKETING ARRANGEMENTS

Market dominated by half a dozen US-based companies who either control or are associated with most of the major producing mines. Much vertical integration.

Index Numbers 1981 = 100

The solid line gives prices in money terms and the dotted line gives prices in 'real' 1981 terms

BARYTES
Drilling mud grade 4·2 SG

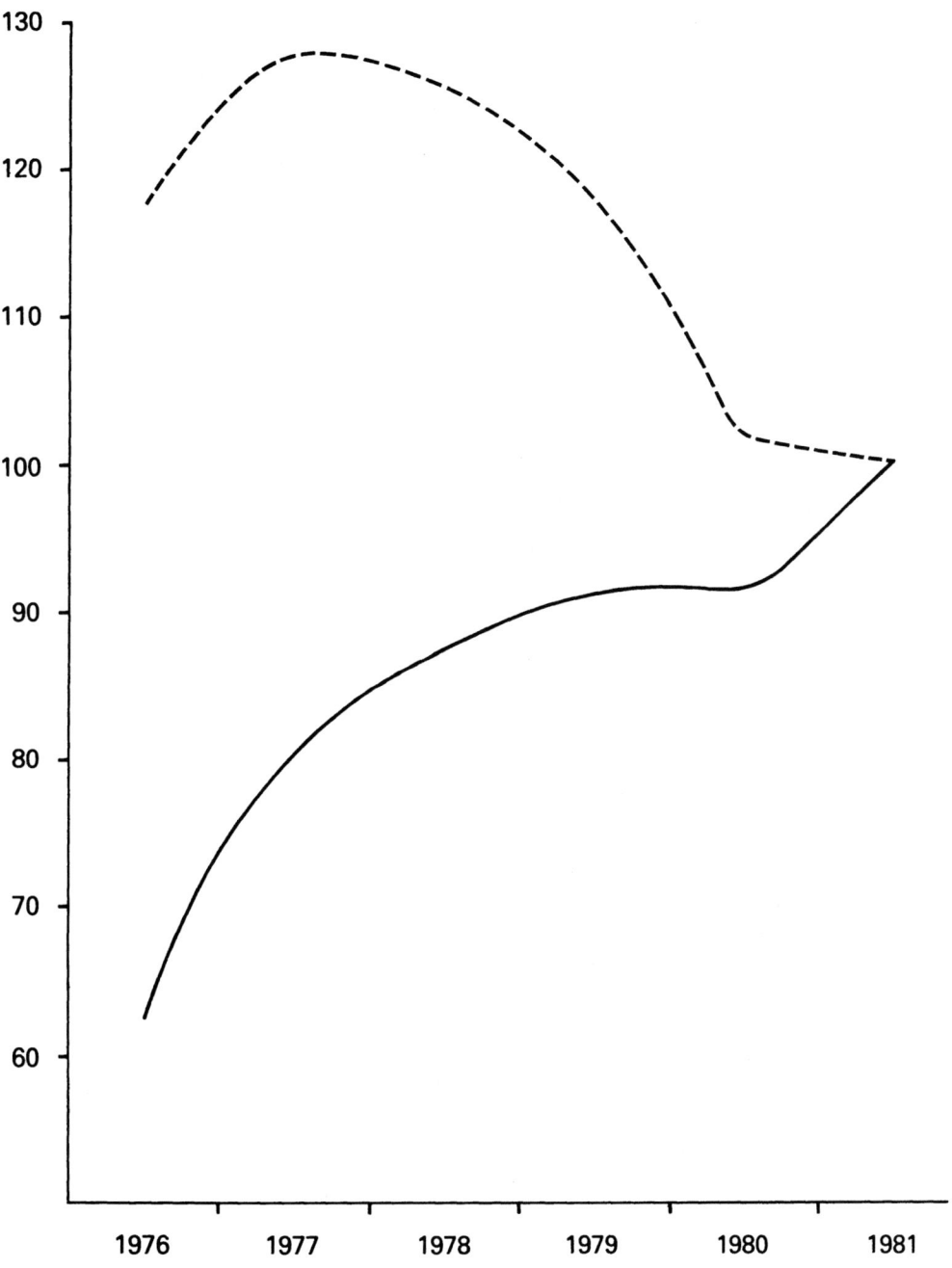

SUPPLY AND DEMAND BY MAIN MARKET AREA

	UK	EC (Ten)	Japan	USA
Production (1979/80 Averages) '000 tonnes	48	1029	56.1	1976
Net Imports (1979/80 Averages) '000 tonnes	116.4	232.7	32.4 (crude & ground)	1514.6 (crude) 13.4 (ground)
Source of Net Imports %				
Australia				1
European Community	80			9
Spain	2	19		
Czechoslovakia		2		
Romania		..		
Chile		6		9
China	3	45	85	23
India				10
Mexico				8
Morocco	12	17		10
Peru				20
Thailand	..	3	15	7
Tunisia	1	1		
Others	2	7		3
Net Exports (1979/80 Averages) '000 tonnes	4.3	425	-	93.4
Consumption (1979/80 Averages) '000 tonnes	160 (apparent)	836.7 (apparent)	90.3	3117
Import Dependence				
Imports as % of consumption	73	28	36	49
Imports as % of consumption and net exports	71	18	36	47
Share of World Consumption % Total World	2	11	1	42
Consumption Growth % p.a. 1970s	6.8	-0.6	0.4	8.9

****

CADMIUM

WORLD RESERVES
('000 tonnes of metal and % of total)

Developed			Less Developed			Centrally Planned			Total
Australia	55	(8.5)	Brazil	40	(6.2)	China	20	(3.1)	
Canada	75	(11.6)	India	10	(1.6)	USSR	45	(7.0)	
Ireland	50	(7.8)	Mexico	20	(3.1)	Others	10	(1.6)	
Japan	20	(3.1)	Namibia	10	(1.6)				
S Africa	50	(7.8)	Peru	35	(5.4)				
Spain	10	(1.6)	Zaire	5	(0.8)				
USA	55	(8.5)	Others	80	(12.4)				
Yugoslavia	10	(1.6)							
Others	45	(7.0)							
Totals	370	(57.4)		200	(31.0)		75	(11.6)	645

These figures are based primarily on estimated world resources of zinc.
World resources on the same basis exceed 1.25 million tonnes. They are
substantially higher when allowance is made for other cadmium-bearing
materials.

WORLD PRODUCTION OF REFINED CADMIUM AT SMELTERS
(tonnes of metal and % of total 1979/80 Averages)

Note: Cadmium is extracted from ores and concentrates, flue dusts and other
materials, which sometimes include scrap. Statistics on mine
production by country are not available.

Developed			Less Developed			Centrally Planned			Total
Australia	882	(4.9)	Algeria	182	(1.0)	Bulgaria	210	(1.2)	
Austria	32	(0.2)	Argentina	20	(0.1)	China	225	(1.2)	
Belgium	1420	(7.8)	Brazil	21	(0.1)	E Germany	15	(0.1)	
Canada	1146	(6.3)	India	121	(0.7)	N Korea	150	(0.8)	
Finland	595	(3.3)	Mexico	825	(4.5)	Poland	747	(4.1)	
France	796	(4.4)	Namibia	86	(0.5)	Romania	90	(0.5)	
W Germany	1233	(6.8)	Peru	190	(1.0)	USSR	2850	(15.7)	
Italy	513	(2.8)	S Korea	50	(0.3)				
Japan	2398	(13.2)	Zaire	206	(1.1)				
Netherlands	408	(2.2)							
Norway	112	(0.6)							
Spain	229	(1.3)							
UK	412	(2.3)							
USA	1701	(9.4)							
Yugoslavia	289	(1.6)							
Totals	12166	(67.0)		1701	(9.4)		4287	(23.6)	18154

RESERVE PRODUCTION RATIOS

Static Reserve life (years)   36
Ratio of identified resources
to cumulative demand 1981-2000   3 : 1
(based on zinc resources alone)

CONSUMPTION

	1979/80 Averages tonnes	% p.a. Growth rates 1970-80
European Community	6549	1.4
Japan	1201	-3.8
United States	4423	-2.1
Others	1524	7.4
Total Western World	13697	-
Total World	17567	0.7

END USE PATTERNS 1980 (USA) %

Transport (including coating & plating of aircraft and vehicles)	17
Coating and Plating	34
Batteries	22
Pigments	13
Plastics and synthetic products	11
Others	3

VALUE OF CONTAINED METAL IN ANNUAL PRODUCTION

$60 million (refined metal at average 1981 prices)

SUBSTITUTES

Zinc can be substituted for some cadmium electroplating applications. Organotin compounds can be used in plastic stabilisers but at higher cost. Cadmium can be substituted in many alloys by a variety of metals, and inorganic compounds can replace it in paints and pigments. Lead-acid batteries can be used as a substitute for nickel-cadmium batteries but at the cost of reliability and longevity.

TECHNICAL POSSIBILITIES

Solar energy cells.

New forms of batteries.

PRICES

	1976	1977	1978	1979	1980	1981
European Free Market: Ingots $1lb	2.7	2.7	2.1	2.5	2.3	1.4
US Producer Metal 99.5% $1lb	2.7	3.0	2.45	2.8	2.8	1.9

Combination of producer and free market prices.

Mainly produced as by-product of zinc smelting and prices tend not to bear relationship to supply/demand balance. Current price levels below cost of production for many producers.

MARKETING ARRANGEMENTS

Environmental pressures becoming an increasingly important restraint on growth especially in industrialised countries. Metal from China becoming more readily available bringing some changes in marketing pattern.

Index Numbers 1981 = 100

The solid line gives prices in money terms and the dotted line gives prices in 'real' 1981 terms

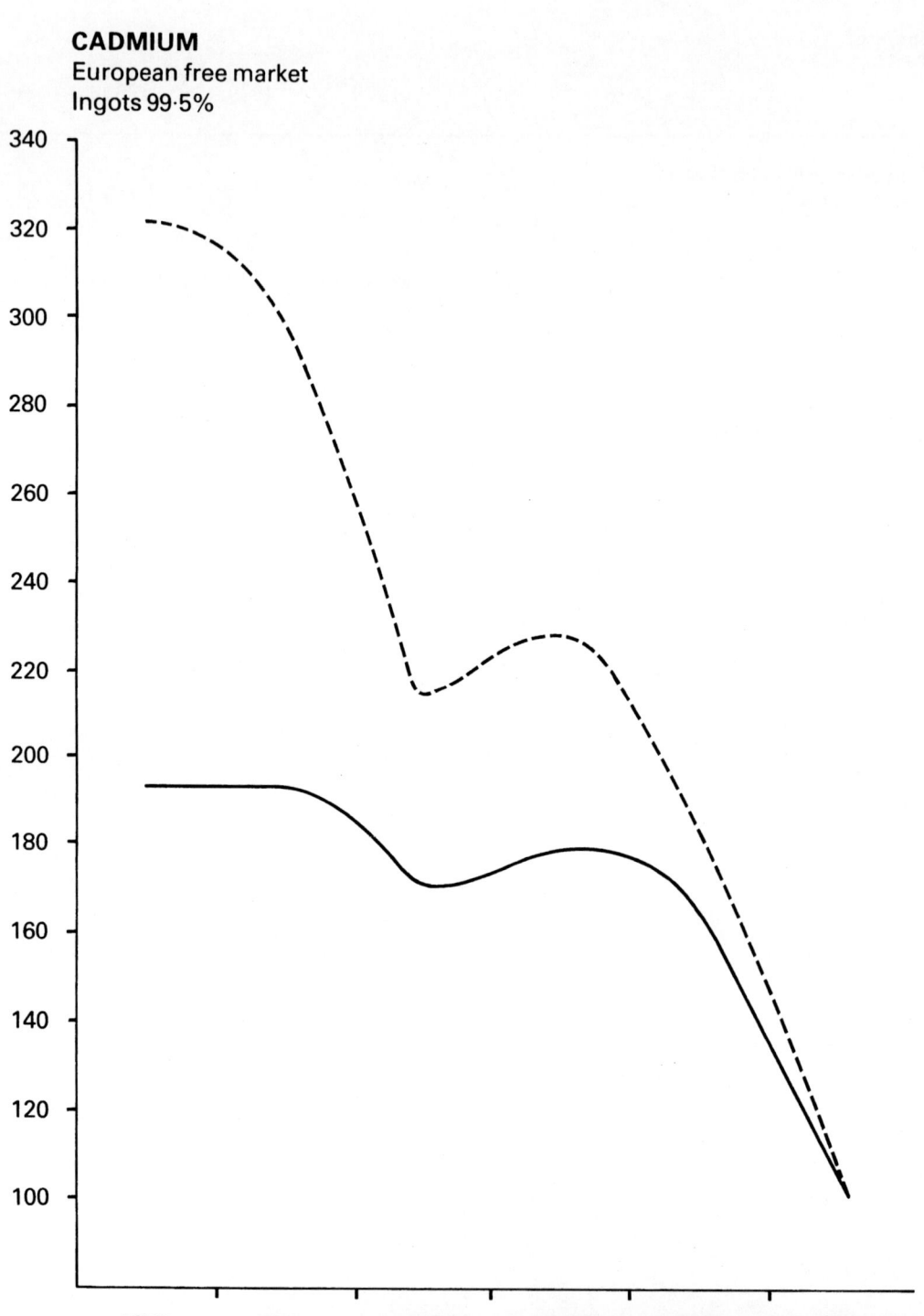

CADMIUM
European free market
Ingots 99·5%

SUPPLY AND DEMAND BY MAIN MARKET AREA

	UK	EC (Ten)	Japan	USA
Production (1979/80 Averages) (tonnes) Refined metal	400	4888	2398	1701

Note: Production in the European Community and Japan is based mainly on imported ores

Net Imports (1979/80 Averages) (tonnes) Refined metal	962	2210	-	2595
Wrought metal	142	110		

Source of Net Imports (%)

	UK	EC (Ten)	Japan	USA
Australia	7	8	-	17
Canada	53	24	-	29
European Community	5	-	-	13
Finland	10	12	-	5
Japan	19	25	-	-
Norway		1	-	3
Spain		3	-	2
S Africa	2	1	-	
Yugoslavia			-	3
Bulgaria		3	-	
China		6	-	
N Korea		3	-	
Mexico		2	-	12
Peru			-	5
S Korea		1	-	7
Zaire		6	-	
Others	4	5	-	4

Net Imports (1979/80 Averages) (tonnes) Refined metal	49	407	190	224
Wrought metal	50	91	-	-

Consumption (1979/80 Averages) (tonnes) Refined metal	1363	6508	1201	4423

Import Dependence

	UK	EC (Ten)	Japan	USA
Imports as % of consumption	71	34	-	59
Imports as % of consumption and net exports	68	32	-	56

Note: For the European Community and Japan this does not take account of the imported raw materials

	UK	EC (Ten)	Japan	USA
Share of World Consumption %				
Western World	10	48	9	32
Total World	8	37	7	25
Consumption Growth % p.a.				
1970s	-0.2	1.4	-3.8	-2.1

****

CHROMIUM

WORLD RESERVES
(million tonnes gross weight and % of total)

Developed			Less Developed			Centrally Planned			Total
Finland	25	(0.8)	Brazil	2	(0.1)	Albania	2	(0.1)	
S Africa	2270	(68.2)	India	5	(0.2)	Cuba	3	(0.1)	
Turkey	5	(0.2)	Iran	2	(0.1)	USSR	15	(0.5)	
			Madagascar	2	(0.1)				
			Philippines	3	(0.1)				
			Zimbabwe	1000	(30.0)				
			Others	1	(..)				
Totals	2300	(69.1)		1015	(30.5)		20	(0.6)	3330

The average chromium content of these reserves is approximately 30%. World Resources total approximately 33,000 million tonnes, gross weight. They are concentrated in South Africa and Zimbabwe (together almost 99% of the total).

WORLD MINE PRODUCTION
('000 tonnes gross weight and % of total 1979/80 Averages)

Developed			Less Developed			Centrally Planned			Total
Cyprus	16	(0.2)	Brazil	345	(3.6)	Albania	1048	(10.9)	
Finland	176	(1.8)	Colombia	5	(0.1)	Cuba	30	(0.3)	
Greece	58	(0.6)	India	314	(3.3)	USSR	2427	(25.2)	
Japan	10	(0.1)	Iran	109	(1.1)	Vietnam	15	(0.2)	
S Africa	3356	(34.8)	Madagascar	139	(1.4)				
Turkey	426	(4.4)	New						
			Caledonia	13	(0.1)				
			Pakistan	3	(..)				
			Philippines	567	(5.9)				
			Sudan	27	(0.3)				
			Zimbabwe	548	(5.7)				
Totals	4042	(42.0)		2070	(21.5)		3520	(36.5)	9632

RESERVE/PRODUCTION RATIOS

Static Reserve life (years)   346
Ratio of identified resources
to cumulative demand
1981-2000   approx.   100 : 1

CONSUMPTION

	1979/80 Averages '000 tonnes	% p.a. Growth rates 1970s
European Community (Ten)	700	6.5
Japan	458	4.1
United States	521 (a)	0.5

The figures cover the chrome content of all forms

(a) Includes 10% secondary

END USE PATTERNS 1980 (USA) %

Chromite: intermediate outlets

Metallurgical industry	57
of which stainless steel	70
Refractory industry	18
Chemical industry	25

Ultimate Markets

Construction	19
Machinery and Equipment	17
Transport	12
Refractories	12
Other	40

VALUE OF ANNUAL PRODUCTION

$1 billion (as chromite at 1981 average prices)

SUBSTITUTES

Substitutes deterred by cost, performance or customer appeal for chromium.

Metals such as nickel, molybdenum and vanadium can be substituted in alloy steels. Zinc, cadmium and nickel can replace chromium for some industrial plating. Titanium is viable in chemical processing equipment and magnesite and zircon for some refractory products. Cadmium yellow is one of several alternative pigments for coating protection.

TECHNICAL POSSIBILITIES

Use of new processes would allow high-iron chromite to be utilised in production of high-carbon ferrochromium.

Increase chromium recovery.

Possibility of semi-continuous stainless steel making would increase chromium use.

PRICES

	1976	1977	1978	1979	1980	1981
Ore Transvaal 44% Cr_2O_3 no ratio $LT	40.2	55.7	56.2	56.0	55.0	53.0
Ore Turkish 48% Cr_2O_3 3:1 ratio $LT	137	137	120.3	108	110	110
Metal US Electrolytic 99.1% Cr $lb	2.5	2.7	2.85	3.2	3.8	4.2
Ferrochrome US Producer Charge 66-70% (3% Si, 5-6.5% C, 44-70% Cr) c/lb	44	42	42	47	50	52

Most ore sold on long term contracts.

MARKETING ARRANGEMENTS

Ore production increasingly highly concentrated, with large state interests - eg: USSR, Etibank in Turkey. Three companies control 70% of South African output and two the Philippines' production. Some ore producers linked with ferroalloy companies such as Union Carbide. Growing trends towards steel industry use of lower grade ferrochrome, and to production of ferrochrome near mines. There is a small free market. Industry affected by problems of steel industry. Strategic value may become of importance in future.

Index Numbers 1981 = 100

The solid line gives prices in money terms and the dotted line gives prices in 'real' 1981 terms

CHROMIUM
Transvaal Ore 44% Cr_2O_3

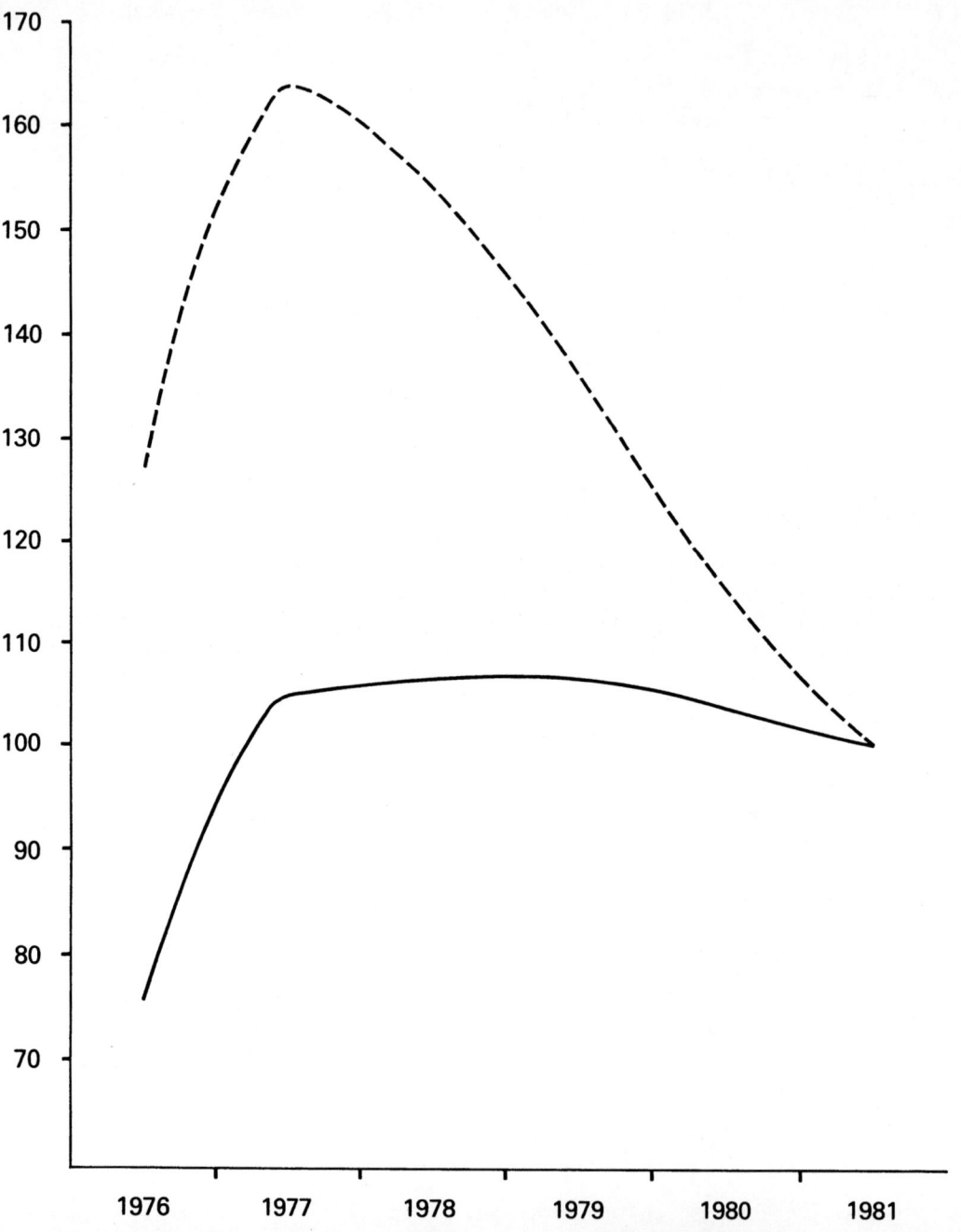

SUPPLY AND DEMAND BY MAIN MARKET AREA

	UK	EC (Ten)	Japan	USA
Production (1979/80 Averages) '000 tonnes				
Chromite Ore (gross)	-	58	10	-
Ferrochromes (gross) and metal	-	159	419	242
		(excl. ore metal)		
Ferrochrome and metal (cr content)	-	97	250	145
		(exc metal)		
Net Imports (1979/80 Averages) '000 tonnes				
Chromite Ore (gross)	92	984	956	910
Ferrochromes (gross)	60	493	277	245
Source of Net Imports (%)				
Chromite				
European Community	3			
Cyprus	1			
Finland	2	2		3
S Africa	72	48	44	41
Turkey		16		8
Albania		17		10
USSR		3		21
India		2	17	
Iran		1		
Madagascar		5		4
Mozambique	9	1		
Philippines	12	2	20	13
Sudan		2		
Others	1	1	19	
Ferrochrome				
European Community	6			2
Finland		2		
Japan				1
Norway	5	1		
S Africa	43	50	80	71
Sweden	28	18	2	5
Turkey	1	6		2
USA			3	
Yugoslavia		3	1	11
USSR	2	1		
Brazil		1	10	2
Zimbabwe		11	2	4
Others	15	7	2	2

	UK	EC (Ten)	Japan	USA
Net Exports (1979/80 Averages) '000 tonnes				
Chromite (gross)	0.6	13.4	–	47.6 (inc. re-exports)
Ferrochrome (gross)	1.0	12.4	14.4	21.3
Consumption (1979/80 Averages) '000 tonnes				
Chromite Ore (gross)	91	1029	757	987
Ferrochromes (gross)	59	641	588	439
Total Consumption (cr content)	70	700	458	521 (inc. c.10% secondary)
Import Dependence				
Imports as % of consumption	100	97	99	c90
Imports as % of consumption and net exports	100	97	99	c90
Share of World Consumption %				
Total World	2	18	12	12 (exc. secondary)
Consumption Growth % p.a.				
1970s	-4.8	6.5	4.1	0.5

****

COBALT

WORLD RESERVES
('000 tonnes of metal and % of total)

Developed			Less Developed			Centrally Planned			Total
Australia	45	(1.5)	Botswana	27	(0.9)	USSR	363	(11.8)	
Canada	27	(0.9)	Morocco	45	(1.5)	Cuba	408	(13.3)	
Finland	18	(0.6)	New						
			Caledonia	91	(3.0)				
USA	318	(10.4)	Philippines	181	(5.9)				
			Zaire	1179	(38.5)				
			Zambia	363	(11.8)				
Totals	408	(13.3)		1886	(61.5)		771	(25.2)	3065

The world's land based resources are estimated at 5.4 million tonnes, and a further 225 million tonnes are contained in seabed nodules.

WORLD MINE PRODUCTION
(tonnes of contained metal and % of total)

Developed			Less Developed			Centrally Planned			Total
Australia	1569	(5.4)	Botswana	292	(1.0)	Cuba	1724	(5.9)	
Canada	1492	(5.1)	Morocco	979	(3.3)	USSR	1928	(6.6)	
Finland	1216	(4.2)	New						
			Caledonia	195	(0.7)				
			Philippines	1255	(4.3)				
			Zaire	15257	(52.1)				
			Zambia	3243	(11.1)				
			Zimbabwe	150	(0.5)				
Totals	4277	(14.6)		21371	(72.9)		3652	(12.5)	29300

WORLD METAL PRODUCTION
(tonnes of metal and % of total)

Developed			Less Developed			Centrally Planned		Total
Canada	472	(1.6)	Zaire	14300	(49.8)	USSR 3605	(12.6)	
Finland	1180	(4.1)	Zambia	3243	(11.3)			
France	725	(2.5)	Zimbabwe	145	(0.5)			
W Germany	343	(1.2)						
Japan	2760	(9.6)						
Norway	953	(3.3)						
UK	535	(1.9)						
USA	437	(1.5)						
Totals	7405	(25.8)		17688	(61.6)	3605	(12.6)	28700

RESERVE/PRODUCTION RATIOS

Static Reserve life (years)	105	(land only)
Ratio of identified resources to cumulative demand 1981-2000	8.5 : 1	(land only)

CONSUMPTION

	1979/80 Averages '000 tonnes	% p.a. Growth rates 1970s
European Community (Ten)	6760	0.2
Japan	2300	0.7
United States	8133	-

The European figures are of dubious reliability, because they are based mainly on trade statistics. They include changes in stocks, which were substantial.

END USE PATTERNS IN 1980 (USA) %

Superalloys	40
Magnetic alloys	15
Cutting & wear-resistant materials	10
Chemical & ceramic use	24
Others (mainly alloy steels, non-ferrous alloys & welding materials)	11

VALUE OF CONTAINED METAL IN ANNUAL PRODUCTION

$1.25 billion (refined metal at 1981 producer prices)

SUBSTITUTES

Trend is towards reduction of, rather than elimination of, cobalt in alloys
eg: iron-base, heat-resistant alloys for cobalt-base materials in turbine
applications.

Ceramic magnets are potential substitutes for permanent magnets. Other
possible future substitutes are molybdenum as a binder for carbides and
nickel, vanadium, chromium or tungsten alloys to replace those containing
cobalt.

Nickel may be substituted for cobalt in several applications but only with a
loss of effectiveness.

TECHNICAL POSSIBILITIES

Exploitation of cobalt-bearing manganese nodules. Recovery of cobalt from
tailings, dumps. Improved scrap recovery.

PRICES

	1976	1977	1978	1979	1980	1981
New York shot/cathode 99.5% Co granules $lb	4.7	5.6	11.4	24.6	25.0	21 *

* 8 months only

Price suspended from September due to confusion in market. Free market
prices dropped below $12/lb by year end.

Mainly produced as by-product of copper or nickel and hence price relatively
independent of supply/demand picture. Mainly producer prices with some free
market. Discounting common in 1981 because of slack demand. Price
susceptible to political disturbances, particularly in Africa.

MARKETING ARRANGEMENTS

Zaire's Gecamines has over half of the market and can strongly influence
price and supply; some scope for varying production even though by-product,
and can stockpile. USSR and Cuba important producers. Canadian nickel
producers make sizeable sales. New nickel sources in Botswana, Australia,
Indonesia and Philippines diffusing sources of supply. Deep sea mining
potential threat to market structure. Strategic metal importance.

Index Numbers 1981 = 100

The solid line gives prices in money terms and the dotted line gives prices in 'real' 1981 terms

COBALT
New York, Shot/Cathode

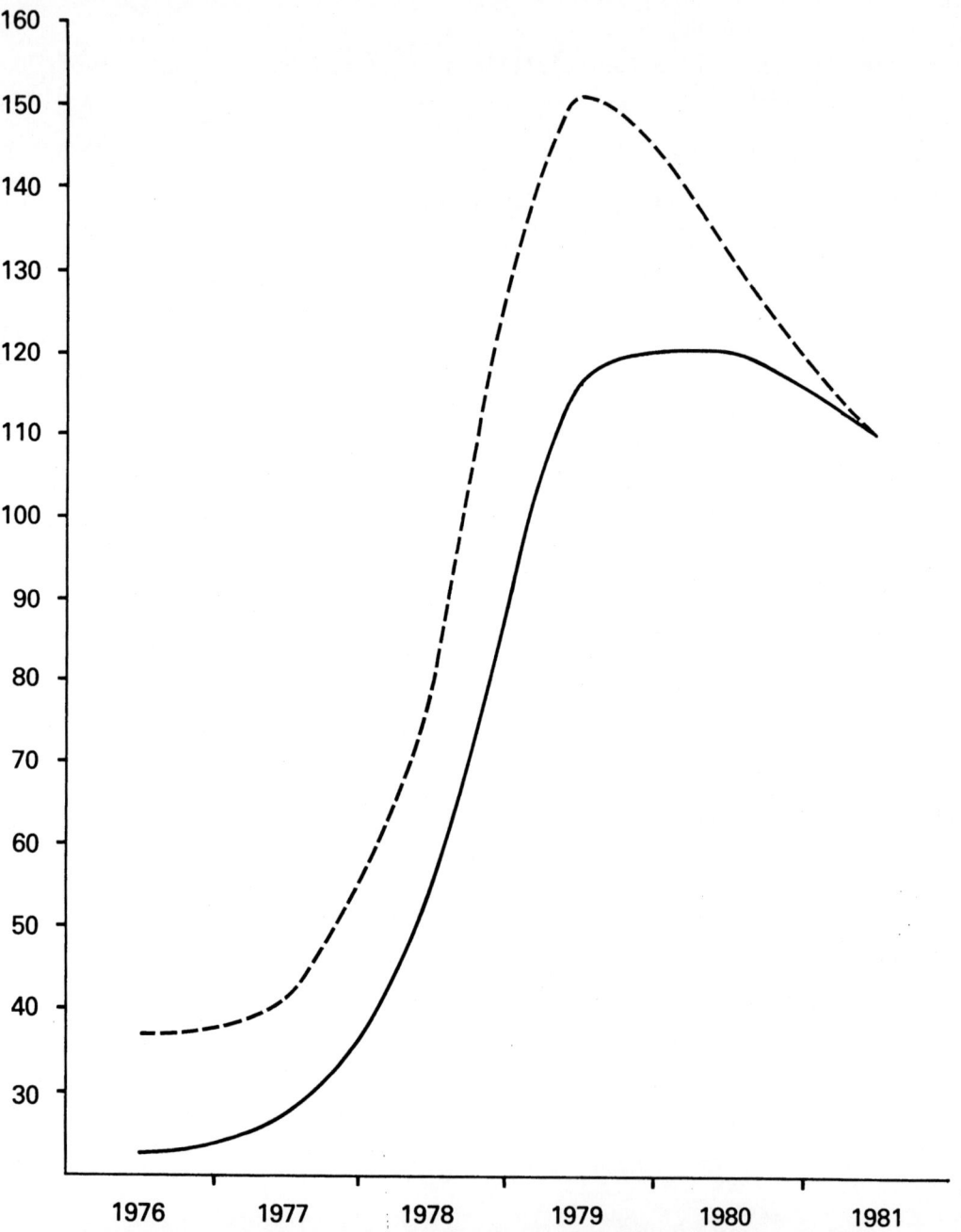

SUPPLY AND DEMAND IN MAIN MARKET AREAS

	UK	EC (Ten)	Japan	USA
Production (1979/80 Averages) (tonnes)				
Mine output	-	-	-	-
Secondary recovery	n/a	n/a	n/a	534
Primary metal	-	1068	2760	437
		(exc. UK & Belgium processing of imported materials)	(from imported area matte from Australia, Philippines & N Caledonia)	
Chemicals				1261
Net Imports (1979/80 Averages) (tonnes)				
Metal	2649	3830	1563	7684
Oxide	1008	1110	471	208
Other forms	n/a	5000	2760	395
		(est. content of ores & partly processed material)	(inc. matte)	(inc. matte)
Total all above forms (Co content)	3340	9650	4660	8233

Source of Net Imports (%)

All forms excluding ores & semi processed cobalt outside USA	UK	EC (Ten)	Japan	USA
Canada	21	20	1	6
European Community	21	-	44	14
(mainly ex Zaire & Canada)				
Finland	2	6	2	6
Japan		7		5
Norway	8	10		6
S Africa				1
Sweden		3		
USA	2	7	14	
Botswana				2
New Caledonia				1
Tanzania		2		

	UK	EC (Ten)	Japan	USA
Zaire	1	12	37	41
Zambia	40	30	1	16
Others	3	5	1	

(Allowing for raw materials Zaire's overall share is c.55%)

Net Exports (1979/80 Averages) (tonnes)

	UK	EC (Ten)	Japan	USA
Metal	941	2428	n/a	297
Oxides	266.5	623.5	47	

Consumption (1979/80 Averages) (tonnes)

	UK	EC (Ten)	Japan	USA
All forms Co content	2500	6760	2300	7421 (reported)

probably
(apparent consumption includes an increase in stocks)

8133 (apparent)

Import Dependence

	UK	EC (Ten)	Japan	USA
Imports as % of consumption (exc.scrap)	100	100	100	100
Imports as % of consumption and net exports (exc. scrap)	100	100	100	100

(94% if scrap included)

Share of World Consumption %

	UK	EC (Ten)	Japan	USA
Total World (approx.)	9	24	8	28

(exc. scrap)

Consumption Growth % p.a.

	UK	EC (Ten)	Japan	USA
1970s	-0.5	0.2	0.7	-

****

COPPER

WORLD RESERVES
(million tonnes and % of total)

Developed			Less Developed			Centrally Planned			Total
Australia	8	(1.6)	Chile	97	(19.7)	Poland 13		(2.6)	
Canada	32	(6.5)	Mexico	25	(5.1)	USSR	36	(7.3)	
S Africa	5	(1.0)	Papua New						
			Guinea	14	(2.8)	Others 11		(2.3)	
USA	92	(18.6)	Peru	32	(6.5)				
Others	19	(3.9)	Philippines	18	(3.7)				
			Zambia	34	(6.9)				
			Zaire	24	(4.9)				
			Others	33	(6.6)				
Totals	156	(31.6)		277	(56.2)		60	(12.2)	493

Total land based resources 1627   ) Total
Deep Sea Nodule resources 690   ) resources 2317

WORLD MINE PRODUCTION
('000 tonnes of combined copper and % of total 1979/80 Averages)

Developed				Less Developed			Centrally Planned			Total
Australia	227		(2.9)	Chile	1064	(13.5)	Bulgaria 61		(0.8)	
Canada	672	(a)	(8.5)	Indonesia	57	(0.7)	China	176	(2.3)	
Finland	39		(0.5)	Mexico	141	(1.8)	Poland	341	(4.3)	
Japan	56		(0.7)	Namibia	41	(0.5)	USSR	1150	(14.6)	
S Africa	208		(2.6)	Papua New						
				Guinea	159	(2.0)	Others	79	(1.0)	
Spain	45		(0.6)	Peru	381	(4.8)				
Sweden	44		(0.6)	Philippines	303	(3.9)				
USA	1306	(a)	(16.7)	Zaire	430	(5.5)				
Yugoslavia	114		(1.4)	Zambia	592	(7.5)				
Others	41		(0.5)	Others	142	(1.8)				
Totals	2752		(35.0)		3310	(42.0)		1807	(23.0)	7869

(a) Averages reduced by strikes in Canada in 1979 and USA in 1980.

WORLD REFINERY PRODUCTION
('000 tonnes metal and % of total 1979/80 Averages)

Developed			Less Developed			Centrally Planned			Total
Australia	177	(1.9)	Brazil	36	(0.4)	Albania	8	(0.1)	
Austria	38	(0.4)	Chile	796	(8.5)	Bulgaria	62	(0.7)	
Belgium	371	(4.0)	Egypt	2	(..)	China	270	(2.9)	
Canada	451	(4.8)	India	16	(0.2)	Czecho-			
						slovakia	25	(0.3)	
Finland	42	(0.4)	Iran	2	(-)	E Germany	51	(0.5)	
France	46	(0.5)	S Korea	80	(0.9)	Hungary	12	(0.1)	
W Germany	378	(4.0)	Mexico	102	(1.1)	N Korea	21	(0.2)	
Italy	14	(0.1)	Peru	230	(2.5)	Poland	347	(3.7)	
Japan	999	(10.7)	Taiwan	17	(0.2)	Romania	66	(0.7)	
Norway	24	(0.3)	Zaire	124	(1.3)	USSR	1465	(15.7)	
Portugal	4	(-)	Zambia	585	(6.3)				
S Africa	150	(1.6)	Zimbabwe	8	(0.1)				
Spain	147	(1.6)							
Sweden	59	(0.6)							
Turkey	21	(0.2)							
UK	142	(1.5)							
USA	1829	(19.6)							
Yugoslavia	134	(1.4)							
Totals	5026	(53.7)		1998	(21.4)		2327	(24.9)	9351

The table includes metal refined from scrap.

SECONDARY PRODUCTION : WESTERN WORLD
('000 tonnes metal 1979/80 Averages)

	Production of Secondary Refined Copper	Direct Scrap used by Manufacturers
European Community (Ten)	356	747
Japan	128	404
USA	470	905
Others	205	431
Total	1159	2487

RESERVE/PRODUCTION RATIOS

Static Reserve life (years)   63
Ratio of identified resources
to cumulative demand 1981-2000   8 : 1
(includes seabed nodules and
 hypothetical and speculative
 resources)

CONSUMPTION OF REFINED METAL

	'000 tonnes 1979/80 averages	Growth Rates % p.a. 1960-70	1970-80
European Community (Ten)	2357	2.3	1.3
Japan	1286	10.4	4.9
USA	2016	4.3	..
Others	1751	4.9	5.0
Total western world	7410	4.3	2.3
Total world	9671	4.4	2.7

END USE PATTERNS 1980 (USA) %

Electrical   58
Construction   18
Industrial Machinery   9
Transport   9
Miscellaneous   6

VALUE OF CONTAINED METAL IN ANNUAL PRODUCTION

$15.7 billion (refined metal at 1981 average price).

SUBSTITUTES

Vulnerable to substitutes on price grounds, technical superiority, or weight both directly (eg: aluminium in electrical uses, optical fibres in telecommunications or plastics in plumbing), or indirectly (eg: aluminium or plastics for brass). Miniaturisation of components also important. Not all substitution is, however, one way; copper can hold its own in many major uses.

TECHNICAL POSSIBILITIES

Possible source in deep sea nodules in 1990s. Expansion of in-situ leaching, and electrochemical processing methods. Expansion of continuous casting of rods direct from cathodes. Uses in solar energy.

PRICES

	1976	1977	1978	1979	1980	1981
Electrolytic wire bar						
US Producer (¢/lb) delivered	68.8	66.8	66.5	93.3	102.4	85.1
LME Cash (¢/lb)	64.0	59.4	61.9	90.1	99.3	79.1
(£/tonne)	782.4	750.25	710.5	934.08	941.75	864.6
LME Range	576.5	638.5	611.5	765	756	777
	to	to	to	to	to	
£/tonne	936.5	903	780.5	1110	1375	981

Most sales are linked to London Metal Exchange or Comex prices which fluctuate markedly. Breakeven costs depend on by-product values, cut-off grades, exchange rates and output. The effective cost floor exceeds ¢75/lb with the cost of significant new capacity substantially over ¢90/lb.

MARKETING ARRANGEMENTS

Over 400 mines but far fewer companies. The US producers who are vertically integrated sell partly on a list price basis. Producer pricing also operates in protected markets such as Japan, South Korea, Taiwan and India, and in major producing countries like Australia, Canada and South Africa. Elsewhere prices are linked to LME quotations, and even in these markets LME exerts major influence. Large mining groups still own large deposits in developed countries, but widespread nationalisation has placed significant portion under state ownership. CIPEC, a government organisation - Chile, Peru, Zaire, Zambia, Indonesia, Mauritania with Papua New Guinea, Yugoslavia and Australia associates - aim to co-ordinate measures to raise copper earnings, but largely ineffective to date in face of over supply, structure of copper industry, and internal conflicts of interest. Strong pressure for an International Copper Agreement, in UNCTAD, but doubtful whether any agreement would work.

Index Numbers 1981 = 100

The solid line gives prices in money terms and the dotted line gives prices in 'real' 1981 terms

COPPER
LME Wirebars, Cash

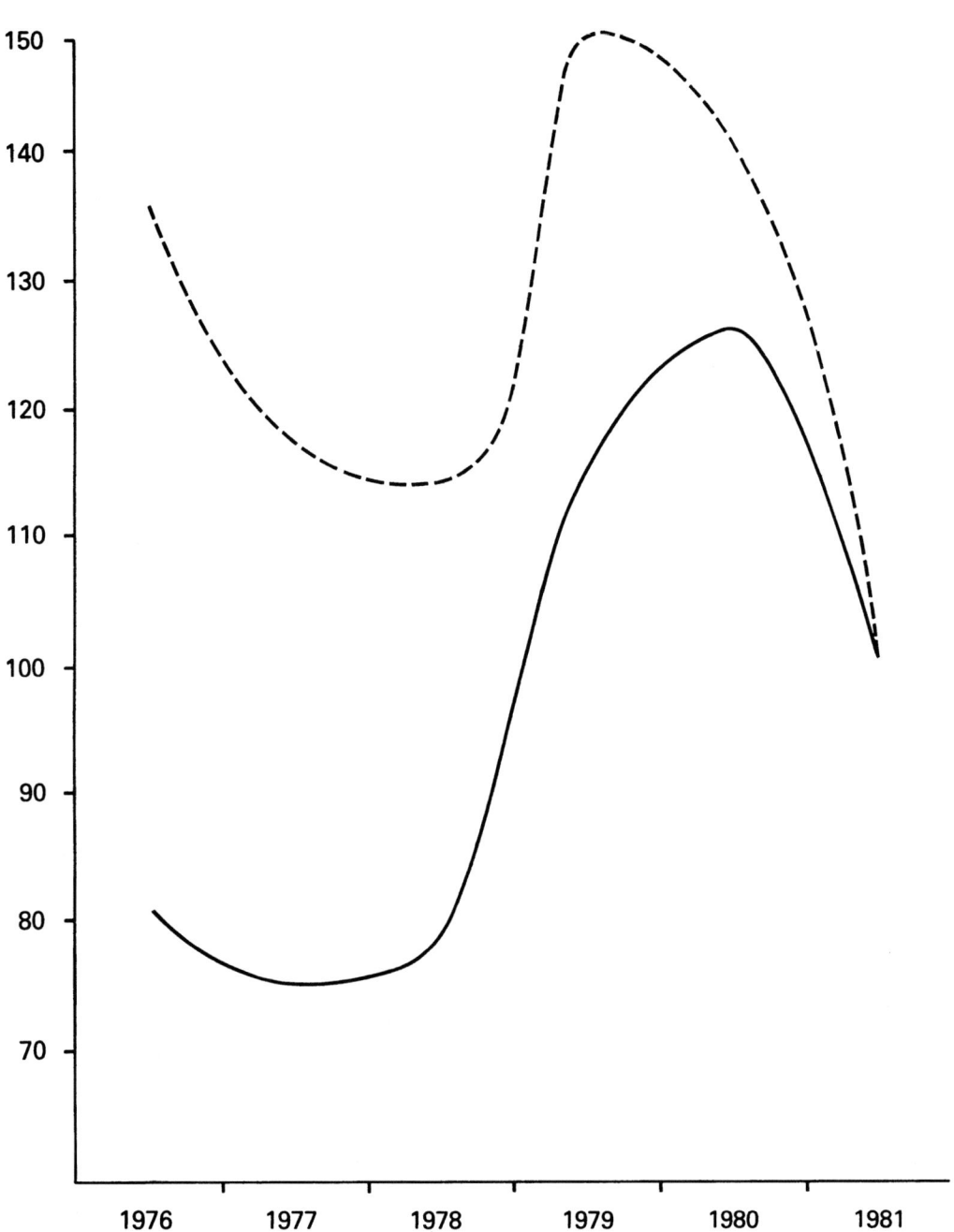

SUPPLY AND DEMAND BY MAIN MARKET AREA

	UK	EC (Ten)	Japan	USA	
Production (1979/80 Averages) ('000 tonnes cu content)					
Mine	0.1	6.8	56.2	1306.0	(a)
Smelter	-	302.7	925.6	1224.6	(a)
of which Secondary	-	146.3	127.4	52.6	(a)
Refined	141.5	950.7	999.0	1833.1	
of which Secondary	83.1	210.2	-	417.5	
Direct Scrap used by Manufacturers	119.5	703 (incomplete)	404.0	883.5	
Net Imports (1979/80 Averages) ('000 tonnes cu content)					
Ores and Concentrates	-	135.6	818.4	25.3	
Blister	67.0	434.0	75.5	57.9	
Refined	294.6	1400.2	266.6	336.7	
Total	361.6	1969.8	1160.5	419.9	

Source of Net Imports (% All Forms)

	UK	EC (Ten)	Japan	USA
Australia	3	3	3	1
Canada	18	8	19	23
European Community	15	-	..	1
Other W Europe	7	8		1
Japan		..	-	16
S Africa - Namibia	2	6	3	..
USA	1	..	6	-
Eastern Bloc	1	6		
Chile	23	24	9	31
Indonesia		1	4	
Malaysia	-	-	2	
Mexico		2	1	1
Papua New Guinea		3	8	
Peru	11	5	7	11
Philippines			20	2
Zaire	1	17	3	1
Zambia	17	16	13	9
Other Countries	1	..	1	3

Note: Details may not add to 100 because of rounding

	UK	EC (Ten)	Japan	USA
Net Exports (1979/80 Averages) '000 tonnes cu content				
Ores and Concentrates	-	4.5	-	75.7
Blister	..	2.7	2.5	5.6
Refined	27.8	101.4	126.3	46.9
Total	27.8	108.6	128.8	128.2
Consumption (1979/80 Averages) '000 tonnes cu content Refined including Secondary but not Direct Scrap	454.0	2356.8	1327.8	2015.5
Import Dependence % Imports as % of Consumption	80	84	87	21
Imports as % of Consumption & Net Exports	75	80	80	20
Share of World Consumption (%) Total Refined:				
Western World	6	32	18	27
Total World	5	24	14	21
Consumption Growth % p.a. 1969/70 to 1979/80	-1.9	1.6	5.0	0.6

Notes: (a)   A strike reduced US output by approximately 300,000 tonnes in 1980, so that these averages are some 150,000 tonnes too low.

****

FLUORSPAR

WORLD RESERVES
(million tonnes contained fluorine in fluorspar and % of total)

Developed			Less Developed			Centrally Planned			Total
Australia	0.18	(0.4)	Argentina	0.73	(1.7)	China	1.72	(4.0)	
Canada	0.82	(1.9)	Brazil	0.82	(1.9)	E Germany	0.64	(1.5)	
France	1.27	(3.0)	India	1.27	(3.0)	Mongolia	1.63	(3.8)	
W Germany	0.54	(1.3)	Kenya	1.18	(2.8)	N Korea	0.09	(0.2)	
Italy	1.0	(2.3)	Mexico	5.44	(12.7)	USSR	2.18	(5.1)	
S Africa	12.70	(29.6)	Morocco	0.54	(1.3)				
Spain	1.45	(3.4)	Namibia	0.91	(2.1)				
UK	3.18	(7.4)	S Korea	0.09	(0.2)				
USA	2.45	(5.7)	Thailand	1.54	(3.6)				
			Tunisia	0.45	(1.1)				
			Zimbabwe	0.09	(0.2)				
Totals	23.59	(55.0)		13.06	(30.4)		6.26	(14.6)	42.91

Figures obtained by multiplying gross fluorspar tonnage by grade of ore and by 0.45. In addition fluorine can be obtained from phosphate rock (fluorapatite). Estimated reserves, assuming that two-thirds of phosphate rock production will be wet processed and that only 22% of contained fluorine will be recovered are 10 million tonnes in the United States and 62 million tonnes in the rest of the world.

World resources are placed at 78 million tonnes of fluorine contained in fluorspar, and 345 million tonnes of fluorine in phosphate rock.

The gross weight of world fluorspar reserves is 560 million tonnes divided as following between the main countries - South Africa 154, USA 129, Mexico 62, United Kingdom 29, France 22, USSR 22, Italy 19, Thailand 19, China 15, Spain 14, Kenya 14, Mongolia 9, other Centrally Planned 6, others 46.

WORLD MINE PRODUCTION : ALL GRADES
('000 tonnes gross weight and % of total 1979/80 Averages)

Developed			Less Developed			Centrally Planned			Total
France	288	(6.2)	Argentina	29	(0.6)	China	399	(8.6)	
W Germany	63	(1.4)	Brazil	75	(1.6)	Czecho-slovakia	96	(2.1)	
Italy	164	(3.5)	India	19	(0.4)	E Germany	100	(2.2)	
S Africa	475	(10.2)	Kenya	100	(2.2)	Mongolia	450	(9.6)	
Spain	267	(5.7)	Mexico	942	(20.3)	N Korea	40	(0.9)	
UK	170	(3.7)	Morocco	59	(1.3)	Romania	20	(0.4)	
USA	92	(2.0)	Thailand	234	(5.0)	USSR	520	(11.1)	
Others	6	(0.1)	Tunisia	34	(0.7)				
			Others	8	(0.2)				
Totals	1525	(32.8)		1500	(32.3)		1625	(34.9)	4650

This table gives gross weights of production regardless of grade. Production of acid and ceramic grade fluorspar included in the above figures, was as follows.

WORLD PRODUCTION OF ACID GRADE FLUORSPAR
('000 tonnes gross weight 1979/80 Averages)

Developed		Less Developed		Centrally Planned		Total
France	167	Argentina	9	Czechoslovakia	48	
W Germany	57	Brazil	37	E Germany	25	
Italy	132	India	11	USSR	249	
S Africa	427	Kenya	89			
Spain	175	Mexico	562			
UK	120	Morocco	59			
USA	47	Thailand	60			
		Tunisia	34			
Totals	1125		861		322	2308
% all grades	74		57		20	50

RESERVE/PRODUCTION RATIOS

Static Reserve life (years) Fluorine in fluorspar   23

Ratio of identified resources to : Fluorine in fluorspar   1.3)

)6.9

Cumulative demand 1981-2000   : Fluorine in phosphate rock 5.6)

CONSUMPTION

	1979/80 Averages '000 tonnes	% p.a. Growth rates 1970s
European Community (Ten)	859	-0.4
Japan	407	-0.5
United States	1168	-2.5

END USE PATTERNS IN USA (1980) %

Steel production	37
Primary aluminium production	40
Chemicals	17
Glass, enamel and other uses	6

VALUE OF ANNUAL PRODUCTION

$0.7 billion (at average 1981 prices).

SUBSTITUTES

Olivine or dolomitic limestone as substitute fluxes in steelmaking.

Gaseous hydrocarbons and carbon dioxide in aerosol propellants.

TECHNICAL POSSIBILITIES

Conservation in the steel industry and recycling and changing technology in the aluminium industry already reducing demand. Environmental concern over fluorocarbons in propellants also restricting demand. However new developments in industrial applications likely to offset these.

PRICES

	1976	1977	1978	1979	1980	1981
US prices, fob Illinois $/net ton						
Metallurgical Pellets 70% CaF_2	89	89.5	90.6	91	98.9	110
Ceramic grade, 88-90% CaF_2	95	95	95	97.5	100	100
Acid grade, 97% CaF_2	105	105	109	115	153.7	165.5

Mainly Producer pricing.

Index Numbers 1981 = 100

The solid line gives prices in money terms and the dotted line gives prices in
'real' 1981 terms

FLUORSPAR
Acid grade 97% CaF₂

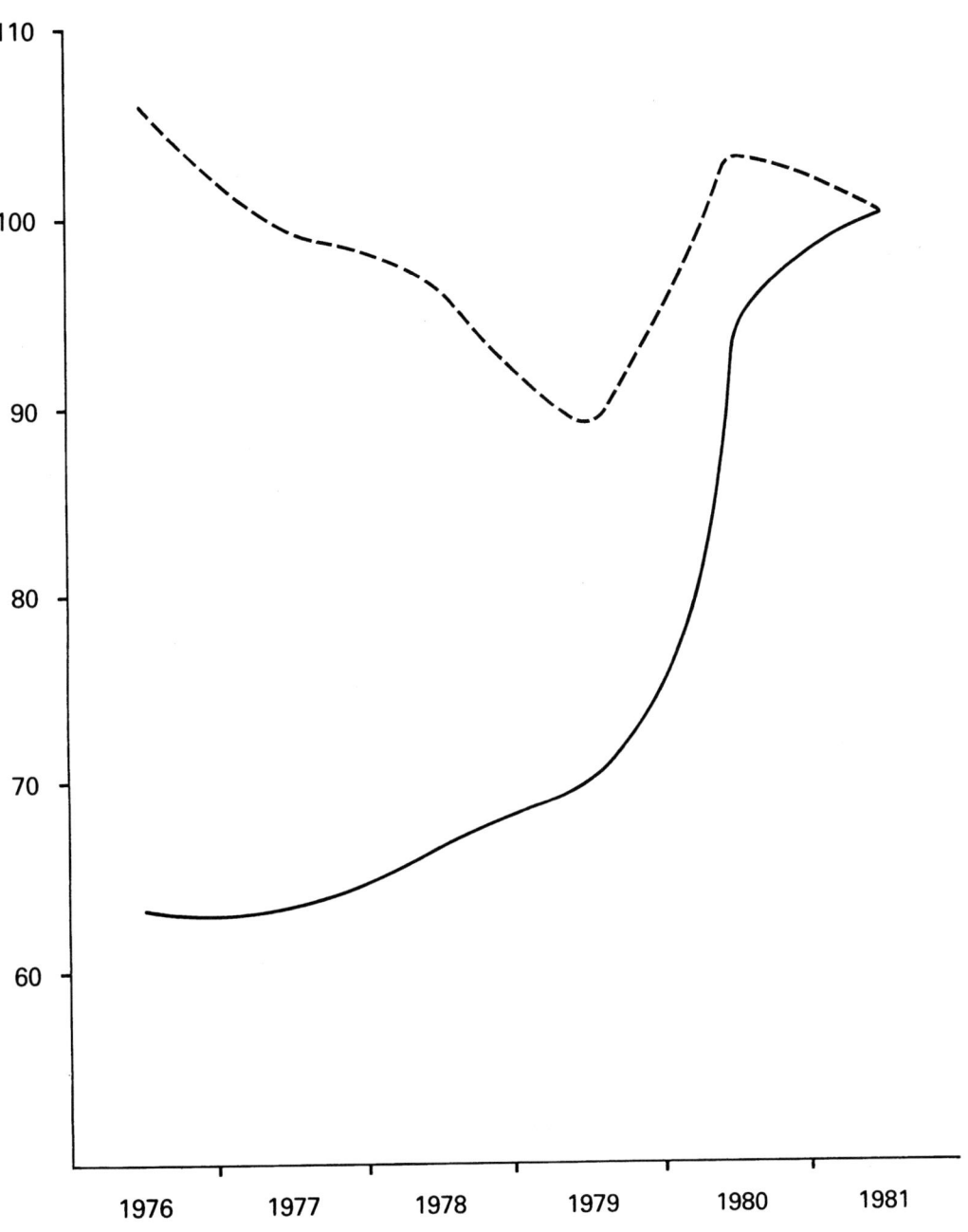

MARKETING ARRANGEMENTS

Substantial vertical integration from main user industries to mines. Small mines in some countries such as Thailand, but industry dominated by large firms. Chinese production has recently been exported to W. Markets.

SUPPLY AND DEMAND IN MAIN MARKET AREA

	UK	EC (Ten)	Japan	USA
Production (1979/80 Averages)				
'000 tonnes gross	170	685	-	92
Fluorspar equivalent from phosphate rock	-		-	66
Net Imports (1979/80 Averages)				
'000 tonnes				
Acid grade more than 97% CaF_2	13.9	108.2	)	(579
			) 477	(
Met grade less than 97% CaF_2	4.9	174.5	)	(292
Fluorspar equivalent from hydrofluoric acid & cryolite				147

Source of Net Imports (%)

Acid Grade			All Grades	
Canada				1
European Community	4			6
S Africa	78	58	26	34
Spain		28		4
China		3	45	
Kenya				3
Mexico				51
Morocco	18	12		
Thailand			27	
Others		4	2	1

Other Grades

European Community	6			
S Africa	41	14		7
Spain		26		
China		16		4
Kenya		11		
Mexico	53	8		88
Morocco		2		
Others		2		1
Undefined		21		

	UK	EC (Ten)	Japan	USA
Net Exports (1979/80 Averages) '000 tonnes				
Acid grade	30.7	83.4	-	
Other grades	11.1	26.0	-	
Total	41.8	109.4	-	14.7
Consumption (1979/80 Averages) '000 tonnes				
Acid grade				553
Other grades				404
All forms (inc. acid etc.)	150	859	407	1168 (apparent)
Import Dependence				
Imports as % of consumption	13	33	100	87
Imports as % of consumption and net exports	10	29	100	86
Share of World Consumption %				
Total World	3	18	9	25
Consumption Growth % p.a.				
1970s	0.2	-0.4	-0.5	-2.5

****

GERMANIUM

WORLD RESERVES

Germanium is obtained as a by-product of zinc or copper-zinc ores. No
reliable data are available for the reserves of large tracts of the world.
The US Bureau of Mines estimates the combined reserves of Canada, the United
States, Europe and Africa at 4,400 tonnes, and their other resources at
4,200 tonnes. The Kipushi mine in Zaire contains the largest known reserve.
Very large potential resources are contained in certain coals, and germanium
might be recovered from ash and flue dusts.

WORLD PRODUCTION

Because of its by-product nature no data are available for mine production
of germanium, but refinery production is estimated as follows:-

(tonnes of contained germanium and % of total 1979/80 Averages)

Developed			Centrally Planned			Total
Belgium	23	(22)	Total	9.5	(9)	
France & Italy	23	(22)				
W Germany	7	(8)				
Japan	15	(15)				
USA	24.5	(24)				
Totals	93.5	(91)		9.5	(9)	103

RESERVE/PRODUCTION RATIOS

Static reserve life (years)   : over 39
Ratio of identified resources
to cumulative demand 1981-2000 : over 3 : 1

CONSUMPTION

	1979/80 Averages '000 tonnes	% p.a. Growth rates 1970s
European Community (Ten)	30 approx.	n/a
Japan	16	2.1
United States	28	4.0

END USE PATTERNS 1980 (USA) %

Electronics & electrical	29
Instruments & optics	66
Others	5

VALUE OF CONTAINED METAL

$100 million (at average 1981 prices)

SUBSTITUTES

Silicon has replaced germanium in some electronic applications but not in high-frequency or high-power applications.

TECHNICAL POSSIBILITIES

Substitute materials could become available for fibre optic applications. Development of superior alternative in some electronic or electrical uses.

PRICES

	1976	1977	1978	1979	1980	1981
$/kg						
US Producer	293	313	318.7	395.6	640.7	879.4
US Dealer	300	330	352.3	471.8	700.2	911.2

By-product of zinc, and certain copper-zinc ores, extracted in refining. Mainly producer priced with dealer market.

MARKETING ARRANGEMENTS

Belgium refines germanium from Zairois ores stockpiled prior to 1975 when Zairois production ceased because of political problems. Production began again in 1980 but currently not being shipped. Residues now coming from US for refining. Producton also in France, Germany, US and Japan. Relatively few producers and consumers and commercial availability governed by rate at which germanium-bearing materials are processed and refined. Some speculative activity.

Index Numbers 1981 = 100

The solid line gives prices in money terms and the dotted line gives prices in 'real' 1981 terms

GERMANIUM
US Producer

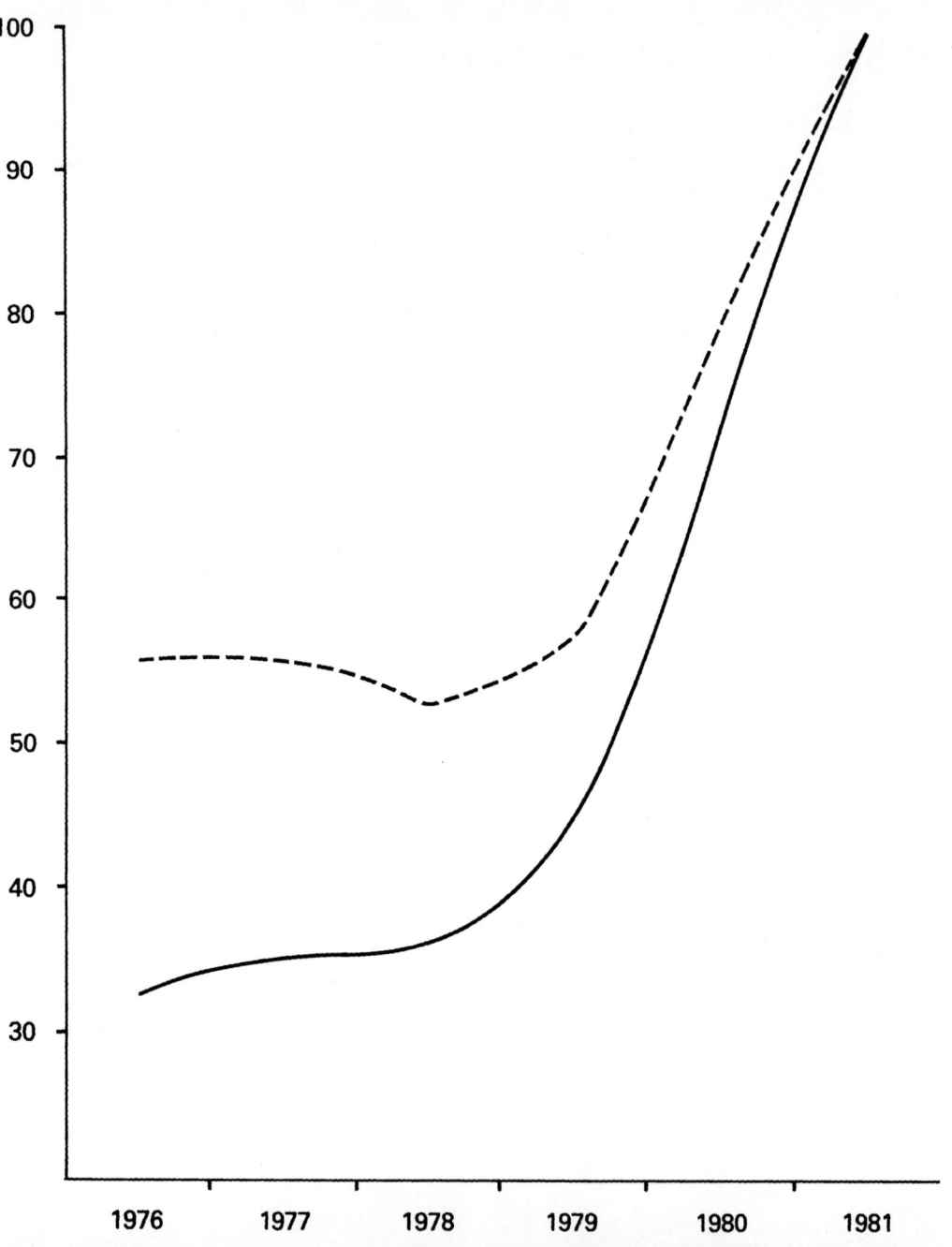

SUPPLY AND DEMAND BY MAIN MARKET AREA

	UK	EC (Ten)	Japan	USA
Production (1979/80 Averages)				
Mine	-	-	-	n/a
Refinery	-	54	15.2	24.5
Net Imports (1979/80 Averages)				
Metal and Oxide	2.5 (metal) large imports of germanium products	1 (metal)	16.7	3.7
Source of Net Imports (%)				
European Community	100		89	98
China		n/a		2
USSR			8	
Others			4	
Net Exports (1979/80 Averages)	1.5 (metal)	25 (based on overseas imports)	-	n/a
Consumption (1979/80 Averages)	1	30	16	28
Import Dependence				
Imports as % of consumption	100	3 (nearer 100 if allowance made for raw materials)	100	13
Imports as % of consumption and net exports	100	2	100	n/a
Share of World Consumption %				
Total World (approx.)	1	30	16	28
Consumption Growth % p.a.				
1970s	n/a	n/a	2.1	4.0

GOLD

WORLD RESERVES
(tonnes of metal and % of total)

Developed			Less Developed			Centrally Planned			Total
Australia	1215	(3.7)	Ghana	125	(0.4)	USSR	7775	(24.1)	
Canada	405	(1.3)	Mexico	375	(1.2)	Others	220	(0.7)	
Japan	155	(0.5)	Philippines	620	(1.9)				
S Africa	16485	(51.1)	Zimbabwe	465	(1.4)				
USA	1400	(4.3)	Other America	1495	(4.6)				
Europe	310	(1.0)	Other Africa	310	(1.0)				
			Pacific	590	(1.8)				
			Other Asia	310	(1.0)				
Totals	19970	(61.9)		4290	(13.3)		7995	(24.8)	32255

Total world resources are estimated at 61,250 tonnes. Because of the
dramatic rise in prices in recent years both this estimate and the figures
in the table are very conservative. In addition above ground stocks of
previously mined gold, held by both central banks and privately, are
substantial.

WORLD MINE PRODUCTION
(tonnes of metal and % of total)

Developed			Less Developed			Centrally Planned			Total
Australia	18.1	(1.5)	Brazil	25.7	(2.1)	China	6.6	(0.5)	
Canada	49.7	(4.1)	Chile	3.5	(0.3)	N Korea	5.0	(0.4)	
Japan	4.0	(0.3)	Colombia	8.5	(0.7)	USSR	256.0	(21.2)	
S Africa	688.7	(57.0)	Dominican Rep.	11.2	(0.9)	Others	3.9	(0.3)	
Spain	3.0	(0.2)	Ghana	12.0	(1.0)				
USA	29.9	(2.5)	Mexico	6.0	(0.5)				
Yugoslavia	4.3	(0.4)	P New Guinea	16.8	(1.4)				
Others	4.9	(0.4)	Peru	4.2	(0.3)				
			Zimbabwe	11.7	(1.0)				
			Others	34.4	(2.8)				
Totals	802.6	(66.4)		134.0	(11.1)		271.5	(22.5)	1208.1

RESERVE/PRODUCTION RATIOS

Based on demand for fabricated gold - i.e. excluding monetary and 'investment' uses. World bullion stocks are ignored. These will make up any shortfalls between mined output and demand.

Static Reserve life (years)   27
Ratio of identified resources
to cumulative demand 1981-2000   1.8 : 1

OVERALL BALANCES OF SUPPLY AND DEMAND IN THE WESTERN WORLD

(tonnes)	1979	1980	1981
Mine Production	960	946	962
Net Trade with Centrally Planned Economies	199	90	283
Net Official Sales	544	-	-
Net Official Purchases	-	-230	-260
Available Supplies	1703	806	985
Fabricated gold in Developed Countries	1060	671	744
Fabricated gold in Less Developed Countries	264	-129	292
Bullion Holdings	379	264	-51

Source: Consolidated Goldfields

INDUSTRIAL USAGE OF GOLD IN THE WESTERN WORLD

	1979/80 Averages '000 tonnes	% p.a. Growth rates 1970s
Jewellery	628.1	-5.7
Electronics	92.3	0.6
Dentistry	74.1	2.2
Other industrial and decorative uses	73.9	1.4
Total	868.4	-4.2
of which European Community (Ten)	316.8	-1.0
Japan	85.7	2.2
United States	157.6	-2.6
Other Countries	308.3	-8.1

Source: Consolidated Goldfields

END USE PATTERNS 1980 (USA) %

Jewellery & arts *	54
Dental	8
Industrial	37
Small items for investment	1
(bars, medallions, coins, etc)	

* Worldwide accounts for 75% of end use; in many LDCs is essentially the only use.

VALUE OF CONTAINED METAL IN ANNUAL PRODUCTION

$18 billion (at average 1981 prices)

SUBSTITUTES

Platinum and palladium substitute to some extent but use influenced by price relationships and established consumer preference for gold. Silver can substitute but is more subject to corrosion. Palladium and bright tin-nickel can be used in electronics. Titanium- and chromium-base alloys are used in dental work.

High prices in 1979-1981 have encouraged substitutes, particularly base metals clad with gold alloy in electronics/electrical industry and in jewellery products.

TECHNICAL POSSIBILITIES

New gold dissolution methods and better media for solvent or resin extraction could improve production technology and utilisation of lower grade sources.

Improvement in recovery rates from industrial waste or scrap.

PRICES

	1976	1977	1978	1979	1980	1981
London fixing am $/troy oz	124.84	147.71	193.36	305.91	614.75	459.7

Prices heavily influenced by speculators in recent years.

MARKETING ARRANGEMENTS

S Africa and USSR produce more than 75% of world's output and are thought to have had talks recently on means of maintaining prices in the face of collapsing demand. State of Russian economy however tends to dictate its sales and IMF auctions plus selling from Central Bank Stockpiles have in the past supplemented supply. Speculative activity particularly in response to political tension can transform market in very short time.

Index Numbers 1981 = 100

The solid line gives prices in money terms and the dotted line gives prices
in 'real' 1981 terms

GOLD
London am fixing

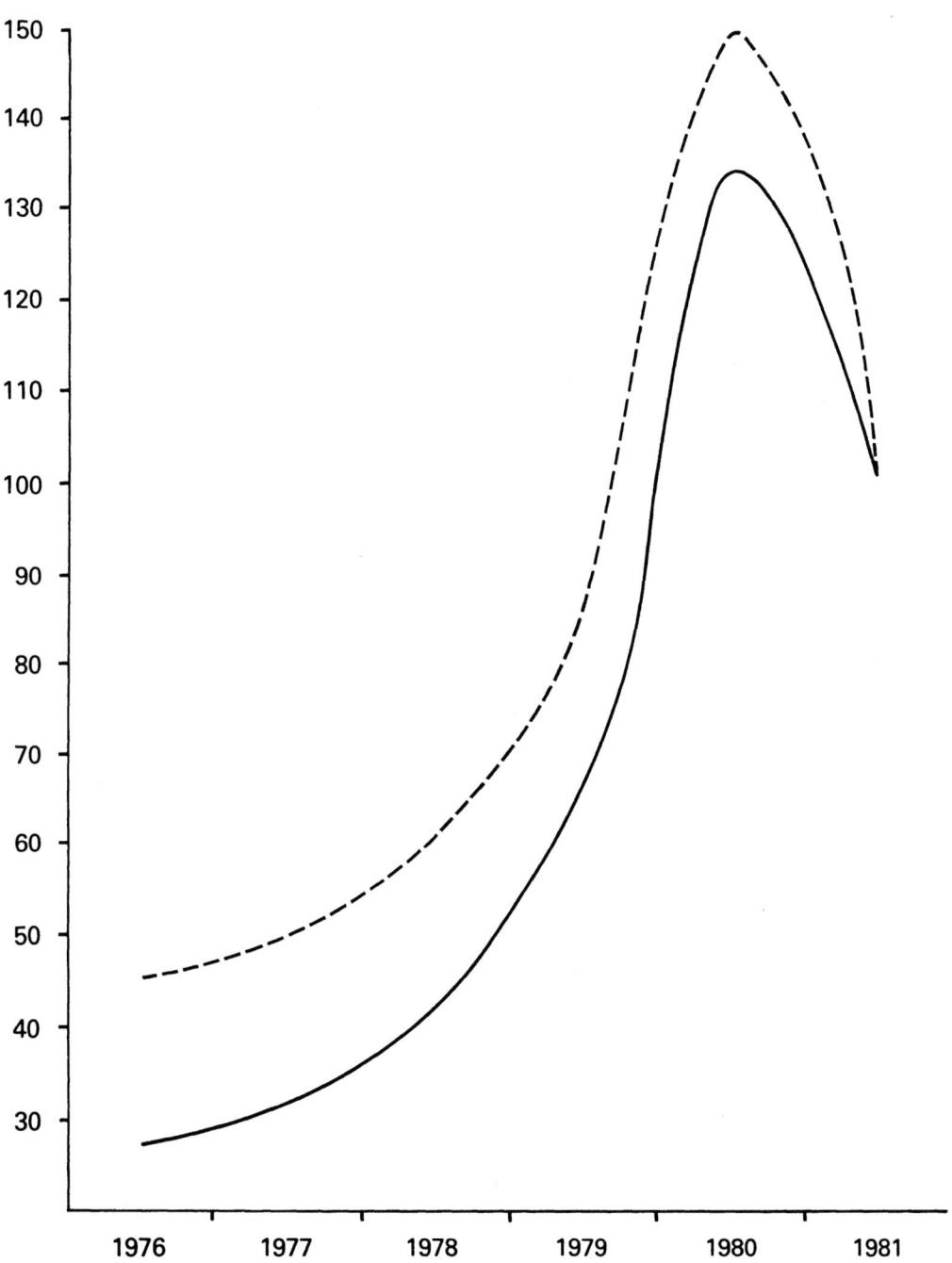

SUPPLY AND DEMAND BY MAIN MARKET AREA

	UK	EC (Ten)	Japan	USA
Production (1979/80 Averages) (tonnes)				
Mine Production	-	1.36	3.58	29.87
Metal (inc. scrap)	n/a	n/a	39.3	168.77
Net Imports (1979/80 Averages) (tonnes)				
Ores and concentrates	8.8	n/a		
From non-ferrous ores and concentrates	-	n/a	28.69	
Unwrought metal	706.3	298.5	40.22)	) 146.64
Wrought metal	76.2 (exc. coin)	42.5	7.13)	(exc. coin)

Note: The European Community figures are from a different source from the UK statistics and do not appear comparable. The UK figures include bullion for monetary transactions

Source of Net Imports (%)

Unwrought metal   n/a

	UK	EC	Japan	USA
Canada			2	45
European Community			46	13
Japan				4
S Africa			31	1
Switzerland			39	44	17
United States			13	
USSR			8	3
Dominican Republic				2
Panama				4
Secret			13	

Wrought metal   n/a   n/a

	Japan
Canada	8
European Community	23
Switzerland	59
USSR	4

	UK	EC (Ten)	Japan	USA

Net Exports (1979/80 Averages)
(tonnes)

Unwrought metal	414.15	20.5	10.05	351.75
Wrought metal	28.95	72.0		

Note: The European Community figures are from a different source from the UK statistics and do not appear comparable. The UK figures include bullion for monetary transactions.

Consumption (1979/80 Averages)
(tonnes)

Reported figures	n/a	n/a	91.2	124.4 (industry & the arts)
Consolidated Goldfields' figures	24.3	31.68	85.7	157.6

Import Dependence

Imports as % of consumption	)	Because of gold's monetary role, its use
Imports as % of consumption and	)	as an investment medium, and the small
net exports	)	share of newly mined output in total
	)	supply, import shares mean very little

Share of World Consumption %
(based on Consolidated Goldfields' figures)

Western world	3	36	10	18

Consumption Growth % p.a.

1970s on reported figures			6.6	-5.3 (industry & the arts)
1970s on Consolidated Goldfields' figures	+0.2	-1.0	2.2	-2.6

****

INDUSTRIAL DIAMONDS

WORLD RESERVES
(million carats and % of total 1979/80 Averages)

Developed			Less Developed			Centrally Planned			Total
Australia	100	(16)	Botswana	125	(20)	USSR	50	(8)	
S Africa	50	(8)	Ghana	15	(2)				
			Zaire	250	(41)				
			Others	30	(5)				
Totals	150	(24)		420	(68)		50	(8)	620

Perhaps 75% of these reserves is in the form of crushing bort with the balance industrial stones. World resources are unknown. Synthetic industrial diamonds supplement reserves.

WORLD MINE PRODUCTION
('000 carats and % of total 1979/80 Averages)

Developed			Less Developed			Centrally Planned			Total
Australia	24	(0.1)	Angola	219	(0.8)	USSR	8550	(29.6)	
S Africa	4982	(17.2)	Botswana	4035	(14.0)				
			Brazil	55	(0.2)				
			Central African Republic	102	(0.4)				
			Ghana	1114	(3.9)				
			Guinea	58	(0.2)				
			Guyana	10	..				
			India	2	..				
			Indonesia	12	..				
			Ivory Coast	33	(0.1)				
			Lesotho	4	..				
			Liberia	131	(0.5)				
			Namibia	80	(0.3)				
			Sierra Leone	443	(1.5)				
			Tanzania	132	(0.5)				
			Venezuela	565	(2.0)				
			Zaire	8370	(28.9)				
Totals	5006	(17.3)		15365	(53.1)		8550	(29.6)	28921

This table does not include substantial illicit production in some less developed countries. For most countries the breakdown between gems and industrial stones has been estimated by the US Bureau of Mines. Apart from the natural diamond covered by the table the USA produced 46 million carats of synthetic diamond, and the rest of the world around 55 million carats.

RESERVE/PRODUCTION RATIOS

Static reserve life (years)   : 21
Ratio of identified resources
to cumulative demand 1981-2000 : under 1, but this
                                 excludes synthetic
                                 diamond and other
                                 resources

CONSUMPTION

	1979/80 Averages '000 carats	% p.a. Growth rates 1970s
European Community	n/a	n/a
Japan	23100	13.9
United States	43050	7.4

The table includes both natural and synthetic diamond.

END USE PATTERNS 1980 (USA) %

Machinery	27
Abrasives	17
Transport equipment	17
Contract construction	13
Stone and ceramic products	13
Mineral services * (drilling bits etc)	6
Other	7

* But accounts for 65% of industrial diamond stone use

VALUE OF ANNUAL PRODUCTION

$150 million (at average US import value of 1981)

SUBSTITUTES

Most substitutes, natural, corundum, and manufactured of fused aluminium oxide, not as efficient or as adaptable. New abrasive materials are being brought into operation and of these cubic boron nitride seems the most promising.

TECHNICAL POSSIBILITIES

Commercial production of large industrial stones suitable for all uses now served by natural stones.

Use of polycrystalline diamond in softer rock formations as substitutes for stones in drilling bits.

PRICES AND MARKETING ARRANGEMENTS

	1976	1977	1978	1979	1980	1981
US Import value for industrial diamonds (excluding diamond dies) $/carat	3.58	3.49	3.96	4.38	5.07	5.34

Most diamond mines produce stones of gem quality and for industrial use.

De Beers Central Selling Organisation controls bulk of world's sales of diamonds of all types although Zaire has recently broken away and now markets independently. Prices vary according to size and grade. The table gives merely a crude indication. Prices of gemstones have risen rapidly in recent years.

Index Numbers 1981 = 100

The solid line gives prices in money terms and the dotted line gives prices in 'real' 1981 terms

INDUSTRIAL DIAMONDS
US Import Value

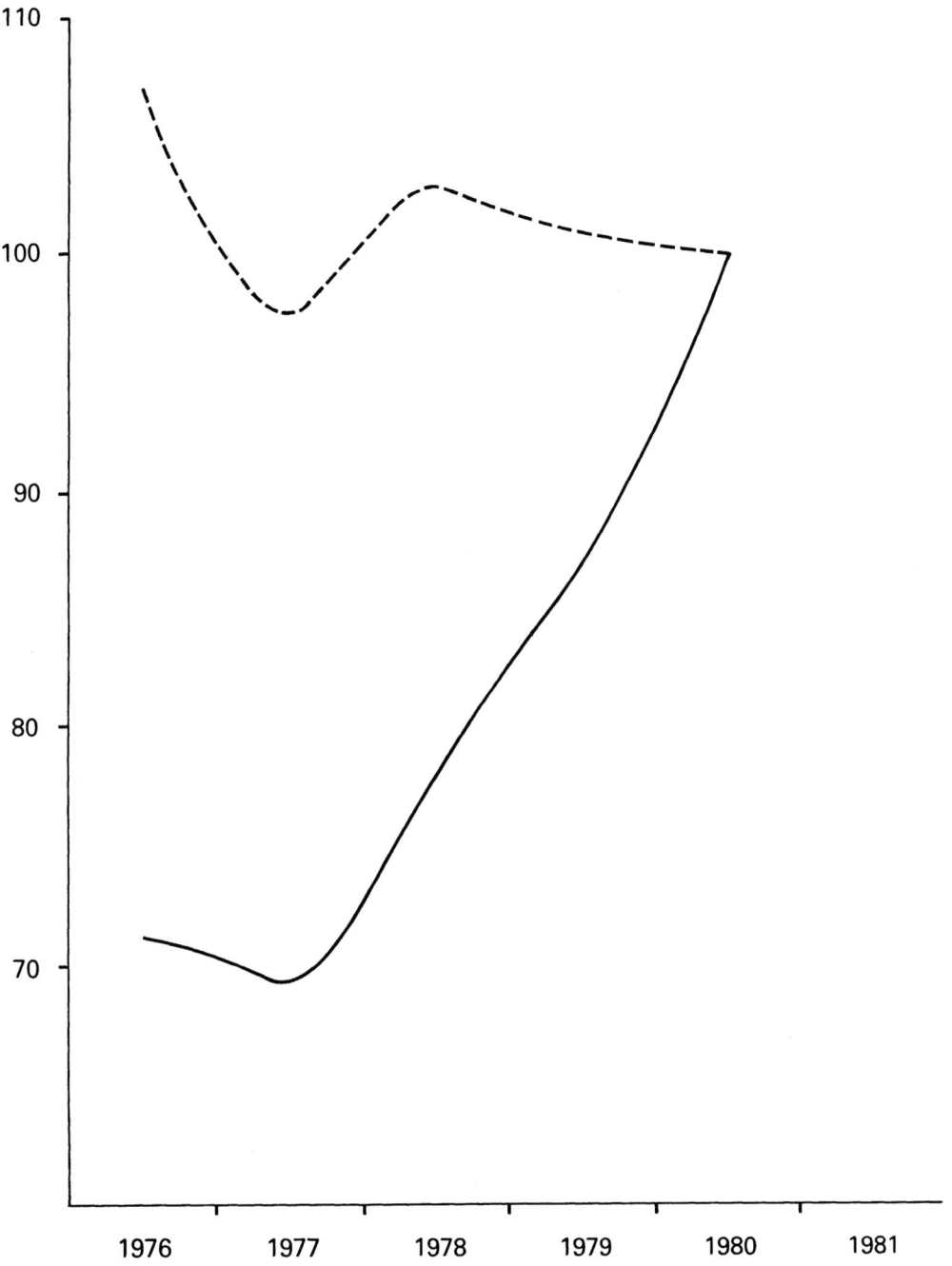

SUPPLY AND DEMAND BY MAIN MARKET AREA

	UK	EC (Ten)	Japan	USA
Production (1979/80 Averages) ('000 carats)				
Natural	-	-	-	-
Synthetic	n/a	n/a	5000	46000
Secondary				3000
Net Imports (1979/80 Averages) ('000 carats)				
Natural	Full figures are		1522	11067
Synthetic	not available		18603	12467
Source of Net Imports (%)				
Natural				
European Community			17	28
S Africa	Details are		11	55
United States	not		22	
Zaire	available		31	11
Others			19	6
Synthetic				
European Community	Details are		41	73
Finland	not			3
Japan	available			6
S Africa				7
Switzerland				4
United States			58	
USSR				4
Others				3
Net Exports (1979/80 Averages) ('000 carats)				
Natural and Synthetic	11229 (1979, natural only)	n/a	-	31050

	UK	EC (Ten)	Japan	USA

Consumption (1979/80 Averages)
('000 carats)

	UK	EC (Ten)	Japan	USA
Natural	n/a	n/a	600)	43050
Synthetic			22500)	

Import Dependence

	UK	EC (Ten)	Japan	USA
Imports as % of consumption	n/a	n/a	80 (all types)	100 (stones) - (Other)
Imports as % of consumption and net exports	n/a	n/a	80 (all types)	100 (stones) - (Other)

Share of World Consumption %

	UK	EC (Ten)	Japan	USA
Total world (approx.) all forms	n/a	n/a	18	33

Consumption Growth % p.a.

	UK	EC (Ten)	Japan	USA
1970s	n/a	n/a	13.9 (all types)	7.4 (all types)

****

IRON ORE

WORLD RESERVES
('000 million tonnes of contained iron and % of total)

Developed			Less Developed			Centrally Planned			Total
Australia	10.7	(11.5)	Brazil	16.3	(17.6)	China	2.7	(2.9)	
Canada	10.9	(11.7)	India	5.6	(6.0)	USSR	28.1	(30.3)	
France	1.6	(1.7)	Liberia	0.6	(0.6)	Others	0.7	(0.8)	
S Africa	1.1	(1.2)	Venezuela	1.3	(1.4)				
Sweden	2.0	(2.2)	Other America	1.5	(1.6)				
USA	3.6	(3.9)	Other Africa	1.5	(1.6)				
Others	2.6	(2.8)	Others	2.0	(2.2)				
Totals	32.5	(35.0)		28.8	(31.0)		31.5	(33.9)	92.8

World reserves amount to roughly 250,000 million tonnes of crude ore giving
the iron contents in the table. Resources exceed 800,000 million tonnes of
crude ore with an iron content over 180,000 million tonnes.

WORLD MINE PRODUCTION
(million tonnes of contained iron and % of total 1979/80 Averages)

Developed			Less Developed			Centrally Planned			Total
Australia	59.2	(11.5)	Algeria	1.9	(0.4)	China	37.5	(7.3)	
Canada	34.3	(6.7)	Brazil	68.2	(13.2)	N Korea	3.1	(0.6)	
France	9.4	(1.8)	Chile	5.0	(1.0)	USSR	131.4	(25.5)	
Norway	2.6	(0.5)	India	25.1	(4.9)	Others	2.2	(0.4)	
S Africa	17.9	(3.5)	Liberia	11.0	(2.1)				
Spain	4.2	(0.8)	Mauritania	5.6	(1.1)				
Sweden	16.8	(3.3)	Mexico	4.5	(0.9)				
USA	49.6	(9.6)	Peru	3.7	(0.7)				
Yugoslavia	1.6	(0.3)	Venezuela	10.0	(1.9)				
Others	7.3	(1.4)	Others	3.3	(0.6)				
Totals	202.9	(39.4)		138.3	(26.8)		174.2	(33.8)	515.4

The gross production of ore from which the above totals were derived
averaged 900 million tonnes. The average grade of ore mined was thus 57.3%.
Average % grades were as follows in 1980 in the leading producing
countries:-

Australia	63		S Africa	64
Brazil	65		Sweden	62.5
Canada	63		USA	63
China	50		USSR	54
India	63		Venezuela	62
Liberia	62			

RESERVE/PRODUCTION RATIOS

Static reserve life (years)   : 186
Ratio of identified resources
to cumulative demand 1981-2000 : 11 : 1

	1979/80 Averages million tonnes fe content	% p.a. Growth rates 1970s
European Community	85	-0.1
Japan	90	7
United States	72	-1.9

END USE PATTERN 1980 (USA) %

Blast furnaces	98.2
Steel furnaces	0.7
Cement production, heavy media materials and others	1.1

VALUE OF CONTAINED METAL IN ANNUAL PRODUCTION

$23.5 billion (at 1981 average prices)

SUBSTITUTES

No substitutes although some scrap is used in steelmaking.

TECHNICAL POSSIBILITIES

Increasing use of direct reduction process is expected to lead to higher steel production in LDCs.

PRICES

	1976	1977	1978	1979	1980	1981
$/tonne						
Brazil 65% fe cif North Sea Ports	22.10	21.59	19.39	23.44	26.70	25.6
Canadian 65% fe cif Germany	20.17	19.74	19.7	24.03	28.94	26.57

Most prices fixed annually under long term sales contracts. Brazil/W Europe price agreement tends to set annual world trend. Influenced by supply/demand conditions in steel industry prevailing at renegotiation and tendency to lag behind economic activity. Freight a major component of prices. Wide price ranges depending on grade and nature of product.

MARKETING ARRANGEMENTS

It is estimated that under 25 companies control almost 80% of world trade. Captive relationships, where steel companies own and operate iron ore mines, are important in US, Canada and Australia especially. Brazil and Australia are most important producers and could work together in future to secure higher prices.

Index Numbers 1981 = 100

The solid line gives prices in money terms and the dotted line gives prices in 'real' 1981 terms

IRON ORE
Brazilian 65% Fe

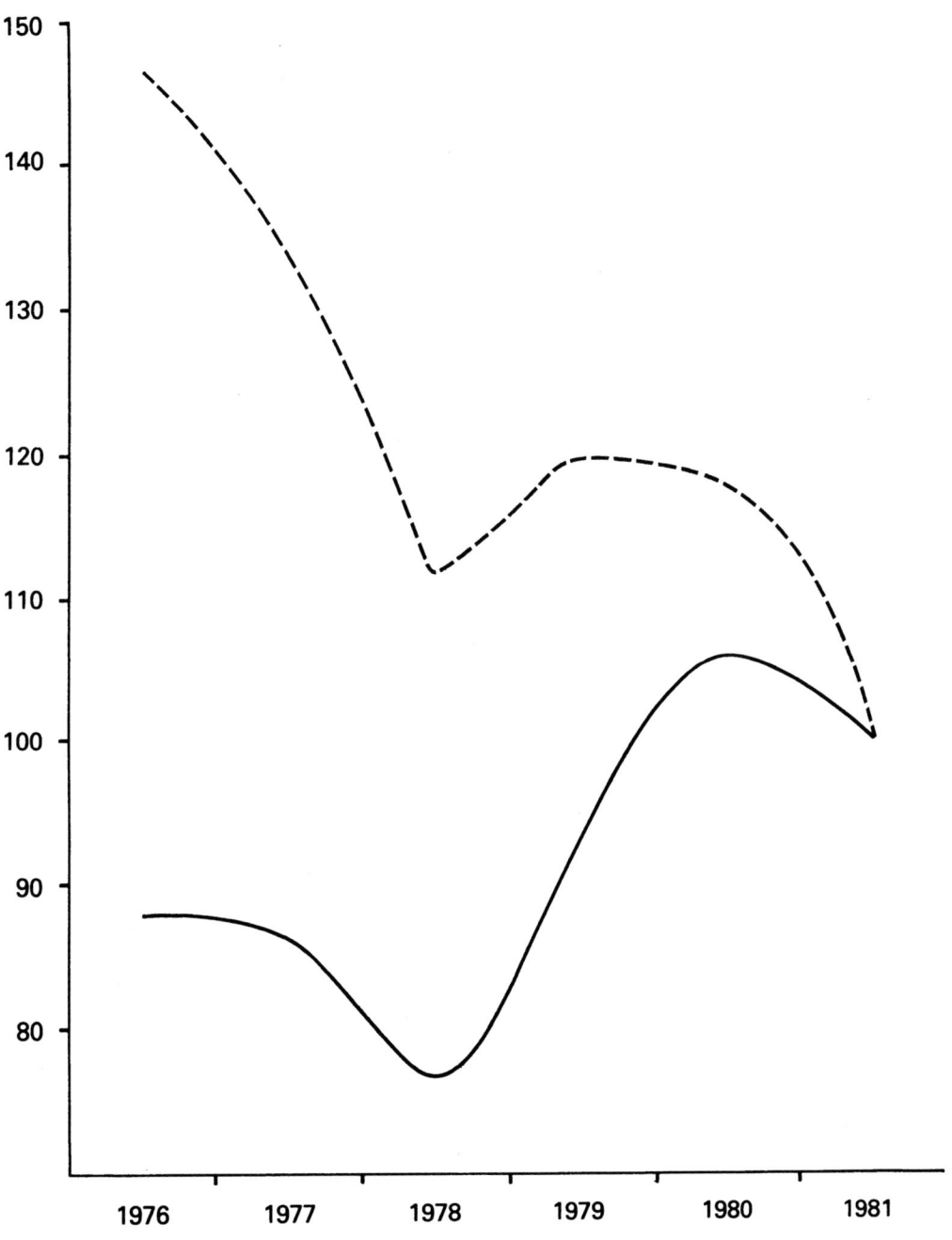

SUPPLY AND DEMAND BY MAIN MARKET AREA

	UK	EC (Ten)	Japan	USA
Production (1979/80 Averages)				
(million tonnes)	2.59	37.91	0.47	78.91
fe content	0.7	11.7	0.24	49.6
Net Imports (1979/80 Averages)				
(million tonnes)	13.19	121.3	126.1	29.89
fe content approx.	8.3	75.9	80.7	18.8
roasted iron pyrites	0.22	0.24		
Source of Net Imports (%)				
Australia	10	9	44	
Canada	37	15	4	68
Norway	9	3		
S Africa	7	6	5	
Sweden	1	15		
Brazil	21	26	21	9
Chile			5	
India			12	
Liberia		12		6
Mauritania	5	6		
Venezuela	7	5		14
Others	3	3	9	3
Net Exports (1979/80 Averages)				
(million tonnes)	-	5.75	-	5.51
roasted iron pyrites		0.06		
Consumption (1979/80 Averages)				
(million tonnes)	15.27 (1980 hit by strike)	150 (approx.)	141.2	113.95
fe content approx.	9.7	85	90	72
Import Dependence (fe content)				
Imports as % of consumption	86	89	89 (most of balance from stock)	26
Imports as % of consumption and net exports	86	87	89 (most of balance from stock)	25

	UK	EC (Ten)	Japan	USA
Share of World Consumption %				
Total world (approx.)	2	16	17	14
Consumption Growth % p.a.				
1970s	-2.2	-0.1	7.0	-1.9

****

LEAD

WORLD RESERVES
(million tonnes of metal and % of total)

Developed			Less Developed			Centrally Planned			Total
Australia	18	(14.2)	Algeria	1	(0.8)	Bulgaria	3	(2.4)	
Canada	12	(9.4)	Brazil	2	(1.6)	China	3	(2.4)	
W Germany	4	(3.1)	Iran	2	(1.6)	Poland	2	(1.6)	
S Africa	5	(3.9)	Mexico	5	(3.9)	USSR	16	(12.6)	
Spain	3	(2.4)	Morocco	1	(0.8)				
Sweden	2	(1.6)	Namibia	1	(0.8)				
USA	27	(21.3)	Peru	3	(2.4)				
Yugoslavia	3	(2.4)	Other America	3	(2.4)				
Others	6	(4.7)	Other Africa	1	(0.8)				
			Other Asia	4	(3.1)				
Totals	80	(63.0)		23	(18.1)		24	(18.9)	127

Total world reserves are estimated at 288 million tonnes.

WORLD MINE PRODUCTION
('000 tonnes of contained metal and % of total 1979/80 Averages)

Developed			Less Developed			Centrally Planned			Total
Australia	408.4	(11.6)	Argentina	34.4	(1.0)	Bulgaria	116.0	(3.3)	
Canada	292.3	(8.3)	Bolivia	16.3	(0.5)	China	155.0	(4.4)	
France	28.9	(0.8)	Brazil	31.8	(0.9)	N Korea	100.0	(2.8)	
W Germany	25.1	(0.7)	Honduras	16.2	(0.5)	Poland	60.4	(1.7)	
Greece	21.8	(0.6)	India	13.0	(0.4)	Romania	33.3	(0.9)	
Greenland	31.1	(0.9)	Iran	15.0	(0.4)	USSR	525.0	(14.9)	
Ireland	65.1	(1.8)	Mexico	159.5	(4.5)	Others	4.7	(0.1)	
Italy	22.0	(0.6)	Morocco	122.9	(3.5)				
Japan	45.8	(1.3)	Namibia	46.0	(1.3)				
S Africa	43.1(a)	(1.2)	Peru	186.7	(5.3)				
Spain	74.6	(2.1)	S Korea	11.3	(0.3)				
Sweden	75.8	(2.2)	Tunisia	10.0	(0.3)				
USA	537.6	(15.3)	Zambia	15.8	(0.4)				
Yugoslavia	124.4	(3.5)	Others	29.9	(0.9)				
Others	21.3	(0.6)							
Totals	1817.3	(51.6)		708.8	(20.2)		994.4	(28.2)	3520.5

(a) In 1980 only from a new mine - i.e. annual production 86

WORLD SMELTER PRODUCTION FROM ORES AND BULLION
('000 tonnes and % of total 1979/80 Averages)

Developed			Less Developed			Centrally Planned			Total
Australia	208	(6.1)	Argentina	29	(0.9)	Bulgaria	105	(3.1)	
Austria	8	(0.2)	Brazil	50	(1.5)	China	148	(4.4)	
Belgium	71	(2.1)	Burma	6	(0.2)	N Korea	62	(1.8)	
Canada	173	(5.1)	India	12	(0.4)	Poland	60	(1.8)	
France	128	(3.8)	S Korea	7	(0.2)	Romania	37	(1.1)	
W Germany	193	(5.7)	Mexico	153	(4.5)	USSR	630	(18.6)	
Greece	20	(0.6)	Morocco	38	(1.1)				
Italy	34	(1.0)	Namibia	42	(1.2)				
Japan	176	(5.2)	Peru	84	(2.5)				
Netherlands	9	(0.3)	Tunisia	18	(0.5)				
Spain	85	(2.5)	Zambia	11	(0.3)				
Sweden	22	(0.6)							
Turkey	4	(0.1)							
UK	119	(3.5)							
USA	563	(16.6)							
Yugoslavia	89	(2.6)							
Totals	1902	(56.0)		450	(13.3)		1042	(30.7)	3394

WORLD REFINED LEAD PRODUCTION
('000 tonnes and % of total 1979/80 Averages)
(This includes secondary antimonial lead)

Developed			Less Developed			Centrally Planned			Total
Australia	245	(4.4)	Argentina	51	(0.9)	Bulgaria	119	(2.1)	
Belgium	99	(1.8)	Brazil	92	(1.7)	China	172	(3.1)	
Canada	244	(4.4)	Mexico	205	(3.7)	E Germany	42	(0.8)	
France	219	(4.0)	Morocco	40	(0.7)	N Korea	68	(1.2)	
W Germany	362	(6.5)	Namibia	42	(0.8)	Poland	83	(1.5)	
Italy	130	(2.3)	Peru	89	(1.6)	Romania	42	(0.8)	
Japan	294	(5.3)	Others	129	(2.3)	USSR	780	(14.1)	
Spain	124	(2.2)				Others	19	(0.3)	
Sweden	45	(0.8)							
UK	347	(6.3)							
USA	1188	(21.4)							
Yugoslavia	106	(1.9)							
Others	163	(2.9)							
Totals	3566	(64.4)		648	(11.7)		1325	(23.9)	5539

LEAD RECOVERED FROM SCRAP : WESTERN WORLD
('000 tonnes 1979/80 Averages)

Scrap included in refined production (i.e. difference between refined and smelter output in previous 2 tables)	1862
Other identified scrap recovery (remelted, alloys and direct use)	256
	2118

RESERVE/PRODUCTION RATIOS

Static reserve life (years)	36
Ratio of identified resources to cumulative demand 1981-2000	2.4 : 1

CONSUMPTION OF REFINED METAL

	1979/80 Averages '000 tonnes	% p.a. Growth rates 1960-70	1970-80
European Community (Ten)	1314	2.4	0.2
Japan	380	8.4	3.4
United States	1220	2.1	0.6
Others	1095	5.8	2.7
Total Western world	4009	3.4	1.2
Total world	5459	3.9	1.8

END USE PATTERNS 1980 (USA) %

Storage batteries	60
Metal products	19
Chemicals (petroleum refining)	12
Pigments (paint, glass ceramics, etc.)	7
Others	2

VALUE OF CONTAINED METAL IN ANNUAL PRODUCTION

$4.3 billion (total refined metal at average 1981 prices)

SUBSTITUTES

Battery replacements include nickel-zinc, zinc-iron and lithium metal-sulphide although large scale commercial use precluded by cost and operating problems.

Polyethylene and other materials substitute in some cable coverings.

In construction applications, plastics, galvanised steel, copper and aluminium are alternatives. In corrosive chemical environments, stainless steel, titanium, plastics and cement are substitutes. Tin, glass, plastics and aluminium are alternatives in tubes and containers, and iron or steel in shot for ammunition.

TECHNICAL POSSIBILITIES

Environmental worries may limit uses for lead particularly in petrol where its use as an anti-knock additive is being phased down. Storage batteries for industrial load levelling, and electric vehicles are prospective markets.

New techniques to recover lead from concentrates and from scrap.

PRICES

	1976	1977	1978	1979	1980	1981
c/lb						
LME Cash	20.2	28.0	29.9	54.6	41.1	33.0
US Producer	23.1	30.7	33.6	52.6	42.4	36.5
£/tonne						
LME Cash	250.7	354.1	342.8	567.7	391.3	362.5
Monthly LME	165.6-	316.2-	298.9-	494.4-	315.6-	293.1-
Range £/tonne	285.5	409.1	432.1	653.7	509.7	452.3

Non US sales based on LME terminal market prices. Substantial percentage of mine output associated with zn, cu, ag, which affects supply and breakeven costs. Large secondary production (with lower costs than primary supply) a major factor.

MARKETING ARRANGEMENTS

Some 300-400 mines mainly as by- or co-product, but smelters the main influence on market trends. Primary smelting dominated by large companies, with state controlled production e.g. from Peru - a growing influence. Secondary smelters, often linked to battery manufacturers, normally have a restraining effect on the market; scrap availability is fairly sensitive to price. There is no international producer agreement.

Index Numbers 1981 = 100

The solid line gives prices in money terms and the dotted line gives prices in 'real' 1981 terms

LEAD
LME Cash

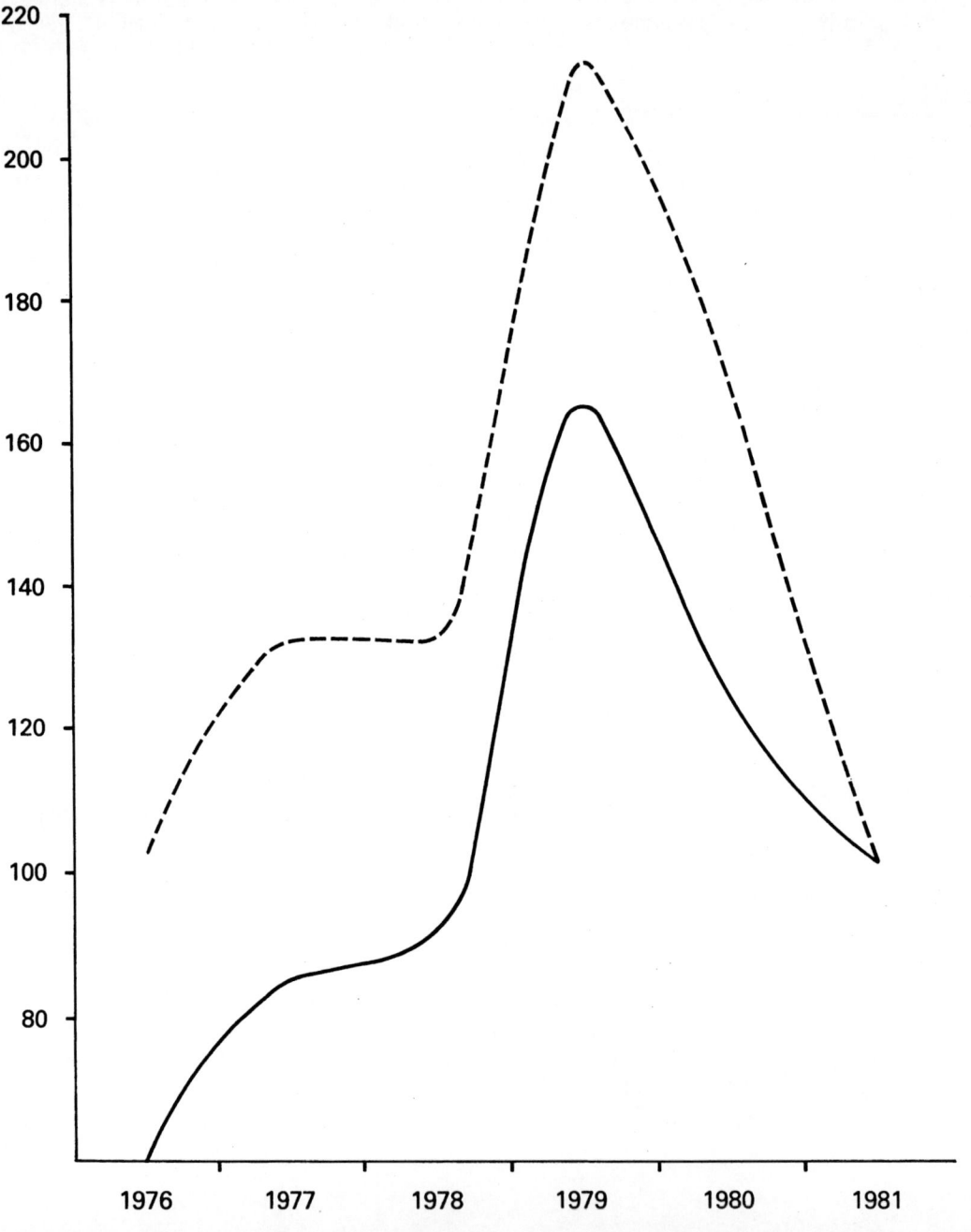

SUPPLY AND DEMAND BY MAIN MARKET AREA

	UK	EC (Ten)	Japan	USA
Production (1979/80 Averages) ('000 tonnes)				
Mine production	2.3	206.9 (inc. Greenland)	45.8	537.6
Smelted from ores and bullion	119	573	176	563
Refined inc. secondary	347	1241	294	1188
Net Imports (1979/80 Averages) ('000 tonnes lead content)				
Ores, flue dust and residues	35.6	258	125.5	37
Base bullion and other unrefined	144.3	208.9)		1
Pigs and bars		)	54.5	
	55.1	187.45		132.5
Reclaimed scrap	10	n/a	3.5	3.4
Source of Net Imports (%)				
Ores etc.				
Australia	52	16	8	8
Canada	15	19	65	15
European Community	8			
Norway	1			
S Africa	4	5	2	
Sweden		14		
Bolivia		5		
Chile				3
Honduras	12	2		26
Iran		1		
Mexico				3
Morocco		17		
Peru	5	9	18	43
S Korea			2	
Thailand		3		
Others	4	8	5	2

	UK	EC (Ten)	Japan	USA
Bullion, Pigs and Bars				
Australia	76	49	17	7
Canada	18	13	2	40
European Community	2			3
S Africa		4		
Sweden		6		
United States		5		
Mexico		4	18	39
Morocco		5		
Namibia			6	1
N Korea		4	26	
Peru		3	20	8
Taiwan			4	
Others	4	7	7	2
Net Exports (1979/80 Averages) '000 tonnes lead content				
Ores and concentrates	2.3	33.4	–	30.3
Lead materials exc. scrap	127	140.6	6.8	87.6
Consumption (1979/80 Averages) ('000 tonnes)				
Refined	314.3	1314	380	1220
Scrap and remelted (not included in refined)	6.3	92	10	112
Import Dependence				
Imports as % of consumption	75	50	47	14
Imports as % of consumption and net exports	53	44	47	13
Share of World Consumption %				
Western world	8	33	9	30
Total world	6	24	7	22
Consumption Growth % p.a.				
1960s	-0.1	2.4	8.4	2.1
1970s	-0.6	0.2	3.4	0.6

LITHIUM

WORLD RESERVES
('000 tonnes lithium and % of total)

Developed		Less Developed		Centrally Planned		Total
Australia	24	Chile	1180	USSR	180	
Canada	180	Zaire	180	Others	5	
USA	415	Zimbabwe	55			
Europe	1	Others	10			
Totals	620 (27.8)		1425 (63.9)		185 (8.3)	2230

Total estimated world resources are approximately 7.6 million tonnes of lithium equivalent.

WORLD PRODUCTION
(tonnes of lithium and % of total 1979/80 Averages)

Developed			Less Developed			Centrally Planned			Total
Portugal	15	(0.2)	Argentina	10	(0.1)	China	355	(4.5)	
USA	5900	(74.6)	Brazil	73	(0.9)	USSR	1250	(15.8)	
			Namibia	48	(0.6)				
			Rwanda	1	(..)				
			Zimbabwe	252	(3.2)				
Totals	5915	(74.8)		384	(4.9)		1605	(20.3)	7904

Note: These figures are estimated from incomplete data. US production is confidential because there are only two producers. The figure above is based on estimated capacity (6300 tonnes of contained lithium), and published US consumption. There is an assumed 15% loss in the production of chemicals from concentrates. Data for other countries are derived from estimated ore production.

RESERVE PRODUCTION RATIOS

Static reserve life (years)   : extremely large
Ratio of identified resources
to cumulative demand 1981-2000   : over 20 : 1

CONSUMPTION

The available statistics are sparse, and those below merely give broad orders of magnitude of consumption of contained lithium as concentrate.

	1979/80 Averages tonnes	% p.a. Growth rates 1970s
European Community	1825	n.a
Japan	510	11.7
United States	3415	5.2

END USE PATTERNS 1980 (USA) %

Primary aluminium	31
Ceramics and glass	28
Lubricants	16
Others	25

VALUE OF ANNUAL PRODUCTION

$125 million (at average 1981 price for lithium carbonate)

SUBSTITUTES

Sodium and potassium substitute as fluxes in ceramics and glass industries. Calcium and aluminium soaps, plus detergents and gels, are alternatives for lithium stearate in lubricants.

Zinc, magnesium, cadmium, sodium and mercury compete for the lithium anode material in batteries. Magnesium has also been successful as a deoxidiser and grain refiner in copper and iron castings.

Lithium can be removed from use in aluminium potlines by increasing the percentages of other salts.

TECHNICAL POSSIBILITIES

Use in nuclear fusion electric power reactors. Development of lithium secondary batteries. Potential for use in the structural metal field and in glass.

PRICES

	1976	1977	1978	1979	1980	1981
US Carbonate 99% min lithium carbonate ¢/lb	83.2	87.3	92.1	106.3	120.3	135.3

MARKETING ARRANGEMENTS

Two US companies control nearly all the Western World's production of lithium concentrate and carbonate. There is limited by-product production elsewhere from brines.

Index Numbers 1981 = 100

The solid line gives prices in money terms and the dotted line gives prices in 'real' 1981 terms

LITHIUM
99·5% Lithium Carbonate

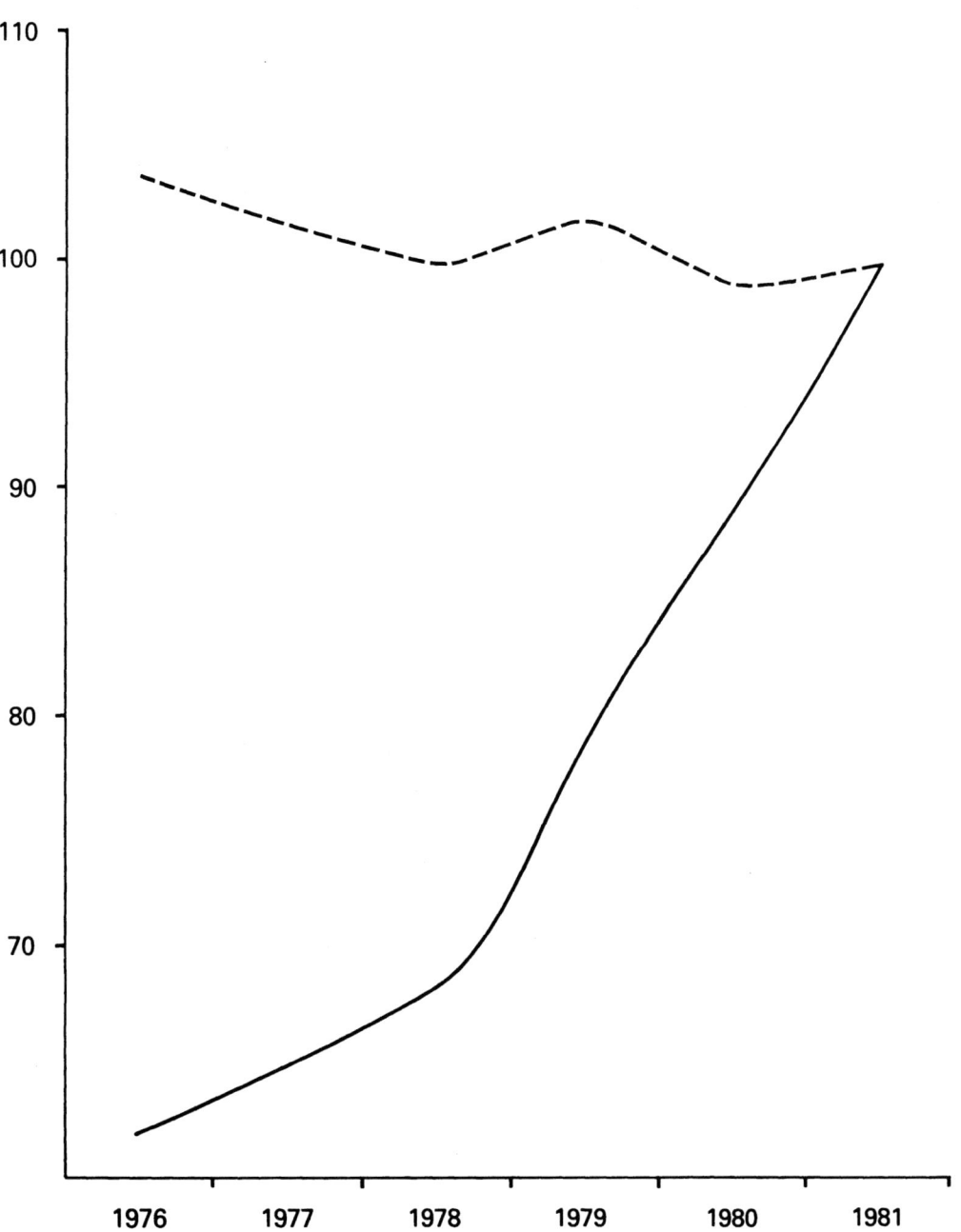

SUPPLY AND DEMAND BY MAIN MARKET AREA

	UK	EC (Ten)	Japan	USA
Production (1979/80 Averages) (tonnes)				
Mine production (contained li)	-	-	-	5900 (estimate)
Products: lithium carbonate	n/a	n/a	n/a	n/a
lithium hydroxide	n/a	n/a	n/a	n/a
Net Imports (1979/80 Averages) (tonnes)				
Ores and concentrates, gross	n/a	c.9500	-	4615
Contained li	n/a	c.300	-	140 (approx.)
Lithium carbonate	415.5	5614.5	2231	40 (all compounds)
Lithium hydroxide	491	1400	691.5	-
Lithium bromide	n/a	n/a	33	-
Lithium metal	4	12	20	-
Total contained lithium	c.165	c.1600	556	c.145 to 150
Source of Net Imports (%)				
Lithium carbonate				
European Community	21			79
United States	70	93	82	
China			13	18
USSR	6	7		
Lithium hydroxide				
European Community	24			
United States	73	77	67	
China	1	2		
USSR	2	17	32	
Lithium bromide				
Israel			37	
United States			62	

	UK	EC (Ten)	Japan	USA
Metal				
European Community	80			
United States		100	100	
Net Exports (1979/80 Averages) (tonnes)				
Lithium carbonate	8	120	-	8881 (all compounds)
Lithium hydroxide	15	93.5	-	2830
Lithium metal	-	12.5	-	n/a
Total all forms contained lithium	c.4	c.50	-	2560 (conc.) or 2177 (chemicals)
Consumption (1979/80 Averages) (tonnes)				
Contained lithium as chemicals after 15% processing losses.	c.160	c.1550	432	2900
i.e. as concentrates	c.190	c.1825	c.510	3415
Import Dependence				
Imports as % of consumption	100	100	100	(net exports)
Imports as % of consumption and net exports	100	100	100	(net exports)
Share of World Consumption %				
Total world (approx.) as concentrate	3	25	7	46
Consumption Growth % p.a.				
1970s	n/a	n/a	11.7	5.2

****

MAGNESIUM

WORLD RESERVES OF MAGNESITE
(million tonnes of magnesium and % of total)

Developed			Less Developed			Centrally Planned			Total
Australia	86	(3.4)	Brazil	136	(5.4)	Czecho-slovakia	18	(0.7)	
Austria	14	(0.5)	India	27	(1.1)	China	745	(29.5)	
Canada	27	(1.1)	Others	325	(12.8)	N Korea	445	(17.6)	
Greece	27	(1.1)				USSR	655	(25.9)	
Turkey	9	(0.3)							
USA	10	(0.4)							
Yugoslavia	5	(0.2)							
Totals	178	(7.0)		488	(19.3)		1863	(73.7)	2529

Identified world resources of magnesite total some 12 billion tonnes. Furthermore magnesium compounds can be recovered economically from well and lake brines and from seawater. The latter which contains 0.13% by weight of magnesium, is a major source of metal and compounds.

WORLD PRODUCTION OF MAGNESITE
('000 tonnes and % of total 1979/80 Averages)

Developed			Less Developed			Centrally Planned			Total
Austria	1100	(10.1)	Brazil	269	(2.5)	Czecho-slovakia	654	(6.0)	
Australia	29	(0.3)	India	378	(3.5)	China	2000	(18.3)	
Canada	56	(0.5)	Mexico	76	(0.7)	N Korea	1837	(16.8)	
Greece	1089	(10.0)	Zimbabwe	85	(0.8)	Poland	19	(0.2)	
S Africa	63	(0.6)	Others	13	(0.1)	USSR	1973	(18.1)	
Spain	386	(3.5)							
Turkey	504	(4.6)							
USA	100	(0.9)							
Yugoslavia	278	(2.5)							
Totals	3605	(33.0)		821	(7.6)		6483	(59.4)	10909

The magnesium content of this production was approximately 3.1 million tonnes. In addition the magnesium content of dolomite, seawater, and well and lake brines amounted to 2.1 to 2.25 million tonnes of contained magnesium, with output in the United States around 853,000 tonnes.

WORLD PRODUCTION OF PRIMARY MAGNESIUM METAL
('000 tonnes and % of total 1979/80 Averages)

Developed			Centrally Planned			Total
Canada	8.95	(2.8)	China	6.5	(2.0)	
France	9.15	(2.9)	Poland	0.5	(0.2)	
Italy	9.25	(2.9)	USSR	75.0	(23.7)	
Japan	10.35	(3.3)				
Norway	44.3	(14.0)				
USA	150.75	(47.7)				
Yugoslavia	1.5	(0.5)				
Totals	234.25	(74.1)		82	(25.9)	316.25

SECONDARY RECOVERY OF MAGNESIUM METAL
('000 tonnes 1979/80 Averages)

Germany	0.7
Japan	20
United Kingdom	2.7
United States	35.2
India	0.1

This includes recovery of magnesium alloys.

RESERVE/PRODUCTION RATIOS

Static reserve life (years) : extremely large (excludes seawater)

Ratio of identified resources to cumulative demand 1981-2000 : over 20 : 1

(This excludes seawater, brines and presently uneconomic resources)

CONSUMPTION OF MAGNESITE

Reliable data for most countries are not readily available. United States' consumption of magnesium compounds averaged 811,000 tonnes of contained magnesium in 1979/80. It fell at an average annual rate of 2% during the 1970s, mainly because of declining steel industry activity.

CONSUMPTION OF MAGNESIUM METAL

	1979/80 Averages '000 tonnes		% p.a. Growth rates 1970s Total
	Primary	Total	
European Community (Ten)	54	54	-2
Japan	19.9	40	8.5
United States	92.8	128	2.8
Other countries	43.3	43	3.2
Total Western world	210	265	3.0
Total world	288	343	3.9

European Community consumption fell in the 1970s because Volkswagen stopped production of its 'Beetle' car which was a heavy user of magnesium metal.

END USE PATTERNS 1980 (USA) %

Non-metal

Refractories	80
Preparation of caustic calcined and specified magnesias and other magnesium compounds	20

Metal

Manufacture of Al based alloys	57
Castings and wrought products	18
Reducing agent	8
Chemicals	7
Nodular iron	4
Other	6

VALUE OF ANNUAL PRODUCTION

Magnesite	$2 billion (at average 1981 prices)
Magnesium metal (primary only)	$900 million (at average 1981 prices)

As magnesite is a raw material for some magnesium metal the two values are not additive.

SUBSTITUTES

Aluminium and zinc are alternatives in many die-casting applications. Sodium can be used to reduce titanium tetrachloride to produce titanium metal.

Rare earth elements and calcium carbide can substitute in production of nodular iron and steel desulphurisation.

Alumina, zirconia, chromite and kyanite are substitutes in magnesia refractories.

TECHNICAL POSSIBILITIES

Development of basic refractories that will withstand high temperature applications, increased use in car industry and in steel desulphurisation.

Development of super-plastic magnesium alloys. Improvement in technology to improve magnesium corrosion.

Olivine and dunite, naturally occurring magnesium compounds, are potential alternatives for silica foundry sand and blasting sand.

PRICES

	1976	1977	1978	1979	1980	1981
Magnesite, dead-burned, bulk, bagged $/st	135	144	150	172	198	218
Magnesium metal US primary ingot 99.8% ¢/lb	89.5	97.5	100.5	105.8	116.7	130.3

MARKETING ARRANGEMENTS

Metal production dominated by US companies but sources of raw materials (sea water, lake brines, magnesite, dolomite) are widespread. Costs of energy a limiting factor on new metal production with present production technology.

Index Numbers 1981 = 100

The solid line gives prices in money terms and the dotted line gives prices in 'real' 1981 terms

MAGNESITE
Dead-burned

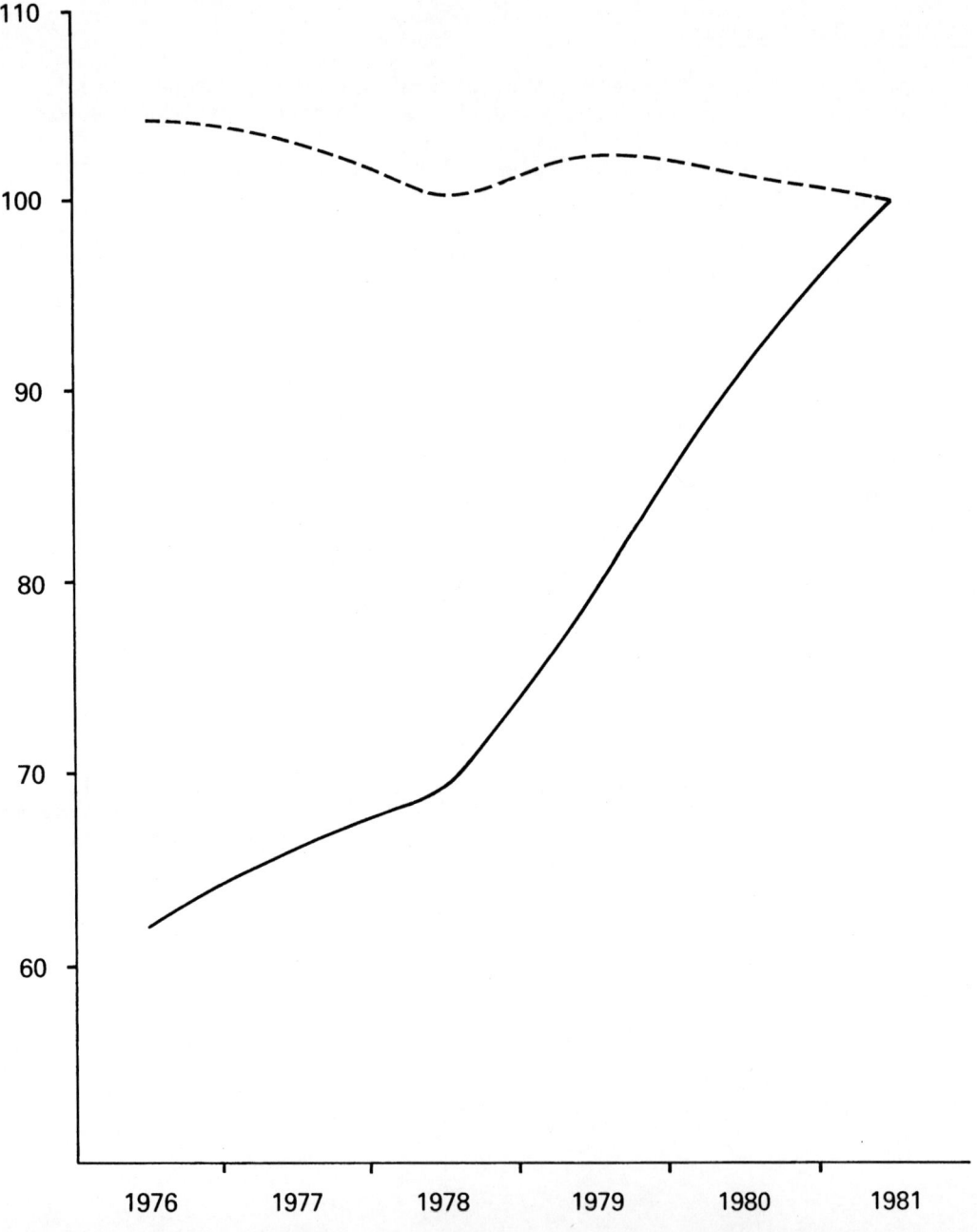

Index Numbers 1981 = 100

The solid line gives prices in money terms and the dotted line gives prices in 'real' 1981 terms

MAGNESIUM
Metal, US Primary Ingot

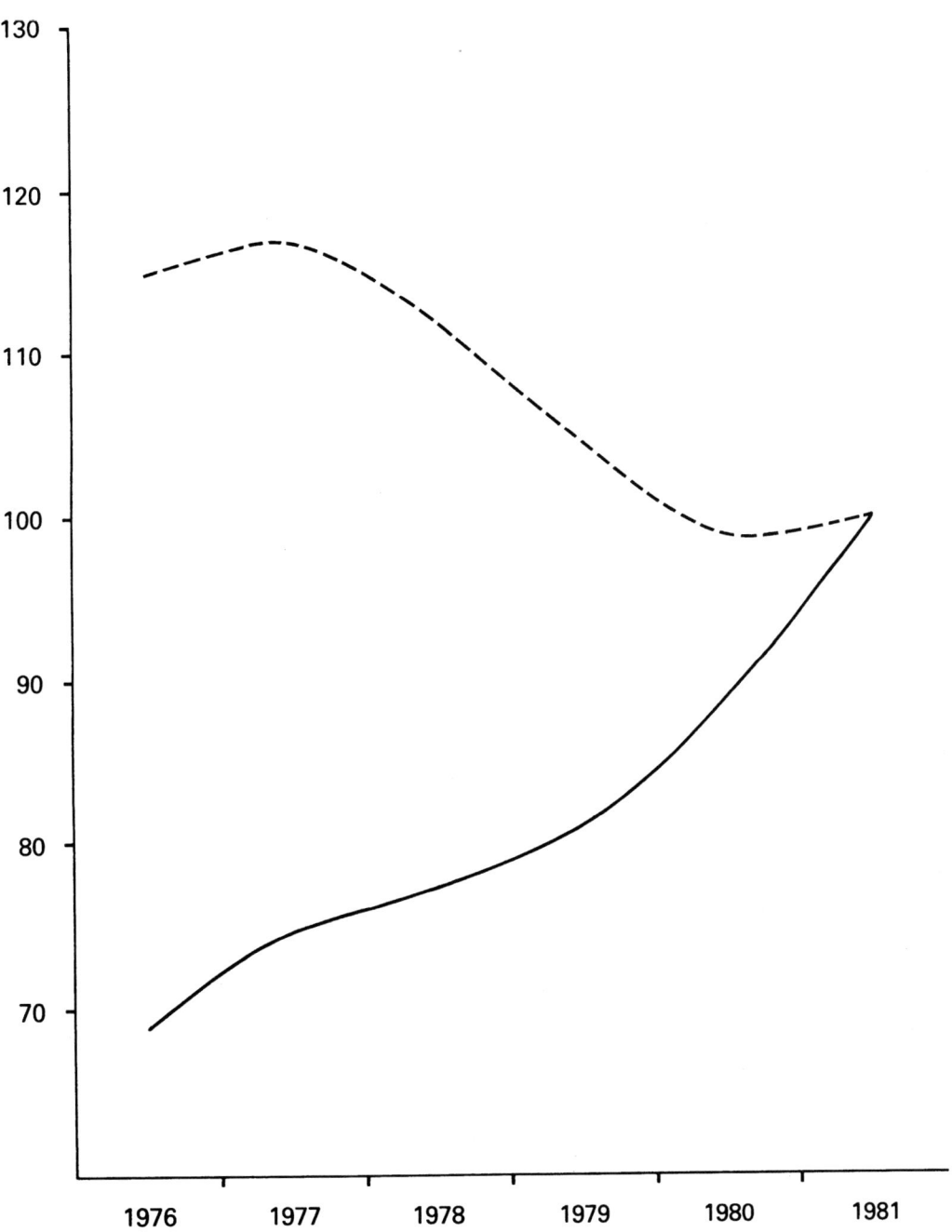

SUPPLY AND DEMAND FOR MAGNESITE BY MAIN MARKET AREA

	UK	EC (Ten)	Japan	USA
Production (1979/80 Averages) '000 tonnes				
Magnesite - gross weight	-	1089	-	c.100
- mg content	-	314	-	c. 17
Magnesia from other sources (dolomite, brines, seawater)				
- mg content	n/a (capacity 220)	n/a	619	853
Net Imports (1979/80 Averages) '000 tonnes				
Magnesium oxide, carbonate and clinker	120	415	600 (clinker 127)	84 (gross) 51 (mg cont)
Source of Net Imports (%) including magnesia from brine and seawater				
Austria	-	20		
European Community	52		42	70
Japan	7	5		18
Spain	22	18		
United States		10	32	
China	9	18	9	
Czechoslovakia		9		
N Korea	3	10	15	
Brazil				4
Israel	3	3		
Net Exports (1979/80 Averages) '000 tonnes	72.5	337	127	95 (gross) 57 (mg cont)
Consumption (1979/80 Averages) '000 tonnes	n/a	n/a	501 (clinker only)	811

	UK	EC (Ten)	Japan	USA
Import Dependence				
Imports as % of consumption	n/a	n/a	25	6
Imports as % of consumption and net exports	n/a	n/a	20	6
Share of World Consumption %				
Total world (approx.)	n/a	n/a	c.11	18
Consumption Growth % p.a.				
1970s	n/a	n/a	n/a	-2

SUPPLY AND DEMAND FOR MAGNESIUM METAL BY MAIN MARKET AREA

	UK	EC (Ten)	Japan	USA
Production (1979/80 Averages) '000 tonnes				
Magnesium metal primary	-	18.4	10.4	150.8
secondary	2.7	3.4	20.2	35.2
Net Imports (1979/80 Averages) '000 tonnes (unwrought and wrought including waste and scrap)	5.87	over 33	12.71	3.86
Source of Net Imports (%)				
Canada	14	3	12	23
European Community	31		3	40
Norway	47	61	14	15
S Africa				5
United States		26	70	
Taiwan				7
Net Exports (1979/80 Averages) '000 tonnes	2.0	3	103.6	50.4
Consumption (1979/80 Averages) '000 tonnes				
(incl. secondary)	5.85	54	40	128
Import Dependence				
Imports as % of consumption	100	over 61	32	3
Imports as % of consumption	75	over 61	9	2
Share of World Consumption %				
Western world	2	20	15	48
Total world	2	16	12	37
Consumption Growth % p.a.				
1970s	-0.5	-2	8.5	2.8

****

MANGANESE

WORLD RESERVES
(million tonnes manganese and % of total)

Developed			Less Developed			Centrally Planned			Total
Australia	118	(8.8)	Brazil	40	(3.0)	Bulgaria	4	(0.3)	
S Africa	717	(53.2)	Gabon	73	(5.4)	China	14	(1.0)	
Others									
(Greece and									
Japan)	1	(0.1)	Ghana	6	(0.4)	USSR	348	(25.8)	
			India	20	(1.5)				
			Zaire	5	(0.4)				
			Others	1	(0.1)				
Totals	836	(62.1)		145	(10.8)		366	(27.2)	1347

The gross weight of these reserves is some 4,500 million tonnes. Land based resources (mainly in South Africa and USSR) amount to 2,800 million tonnes of contained manganese. In addition sea bed nodules contain an estimated 16,000 million tonnes of manganese.

WORLD MINE PRODUCTION
('000 tonnes gross weight and % of total 1979/80 Averages)

Developed			Less Developed			Centrally Planned			Total
Australia	1814	(6.9)	Brazil	2218	(8.4)	China	1540	(5.8)	
S Africa	5439	(20.6)	Gabon	2223	(8.4)	USSR	10250	(38.7)	
Others	152	(0.6)	Ghana	262	(1.0)	Others	126	(0.5)	
			India	1700	(6.4)				
			Mexico	470	(1.8)				
			Morocco	142	(0.5)				
			Others	114	(0.4)				
Totals	7405	(28.0)		7129	(27.0)		11916	(45.0)	26450

The table excludes modest production of low grade ore in several countries. It averaged 120 to 150,000 tonnes gross weight in 1979/80.

The manganese content of mined ore varies widely between countries. The overall manganese content of world mine output averaged approximately 10,250 million tonnes in 1979/80. The shipped ore grades of the main producers (% of contained manganese) are:-

Australia	37-53
Brazil	38-50
China	20+
Gabon	50-53
Ghana	30-50
India	10-54
Mexico	35+
Morocco	50-53
S Africa	30-48+
USSR	35

RESERVE/PRODUCTION RATIOS

Static reserve life (years)	131
Ratio of identified resources to cumulative demand 1981-2000	
Land Based resources only	10 : 1
Land and Sea Bed resources	70 : 1

CONSUMPTION

	1979/80 Averages '000 tonnes	% p.a. Growth rates 1970s
Manganese ore (gross weight)		
European Community	3051	-0.2
Japan	1521	1.5
United States	1108	-6.0
Ferro Manganese (gross weight)		
European Community	1031	0.5
Japan	812	2.1
United States	801	-1.6

Note:- The ferro manganese figures in this table and in the table on supply and demand by main market area include some double counting of high carbon ferro manganese that is used to make more refined products.

END USE PATTERNS 1980 (USA) %

Construction	23
Transport	20
Machinery	16
Other (including chemicals, batteries, oil and gas industry)	41

VALUE OF ANNUAL PRODUCTION

$1.8 billion (metal content at average 1981 prices)

SUBSTITUTES

No substitutes in major applications.

TECHNICAL POSSIBILITIES

Deep sea nodules.

PRICES

	1976	1977	1978	1979	1980	1981
Ore						
Europe 48-50% Mn $/tonne of contained metal	144	148	140	135	164	172
Metal						
UK Electrolytic min 99.95% £/tonne	600.4	660	615.4	623.3	670.0	700.6

Prices negotiated, dependent on chemical quality, physical character, quantity, delivery terms etc. Published quotations only reflect general condition of market. Freight charges particularly important.

MARKETING ARRANGEMENTS

A few large companies dominate, with government ownership important in some cases. Some steel producers (e.g. Bethlehem and BHP) have manganese interests. Five countries control approximately half of non Eastern Bloc ore production with South Africa dominating. Trend to forward integration by ore producers into ferro manganese production - e.g. in South Africa. Much ore trade handled by agents.

Index Numbers 1981 = 100

The solid line gives prices in money terms and the dotted line gives prices in 'real' 1981 terms

MANGANESE
Ore, 48–50% Mn

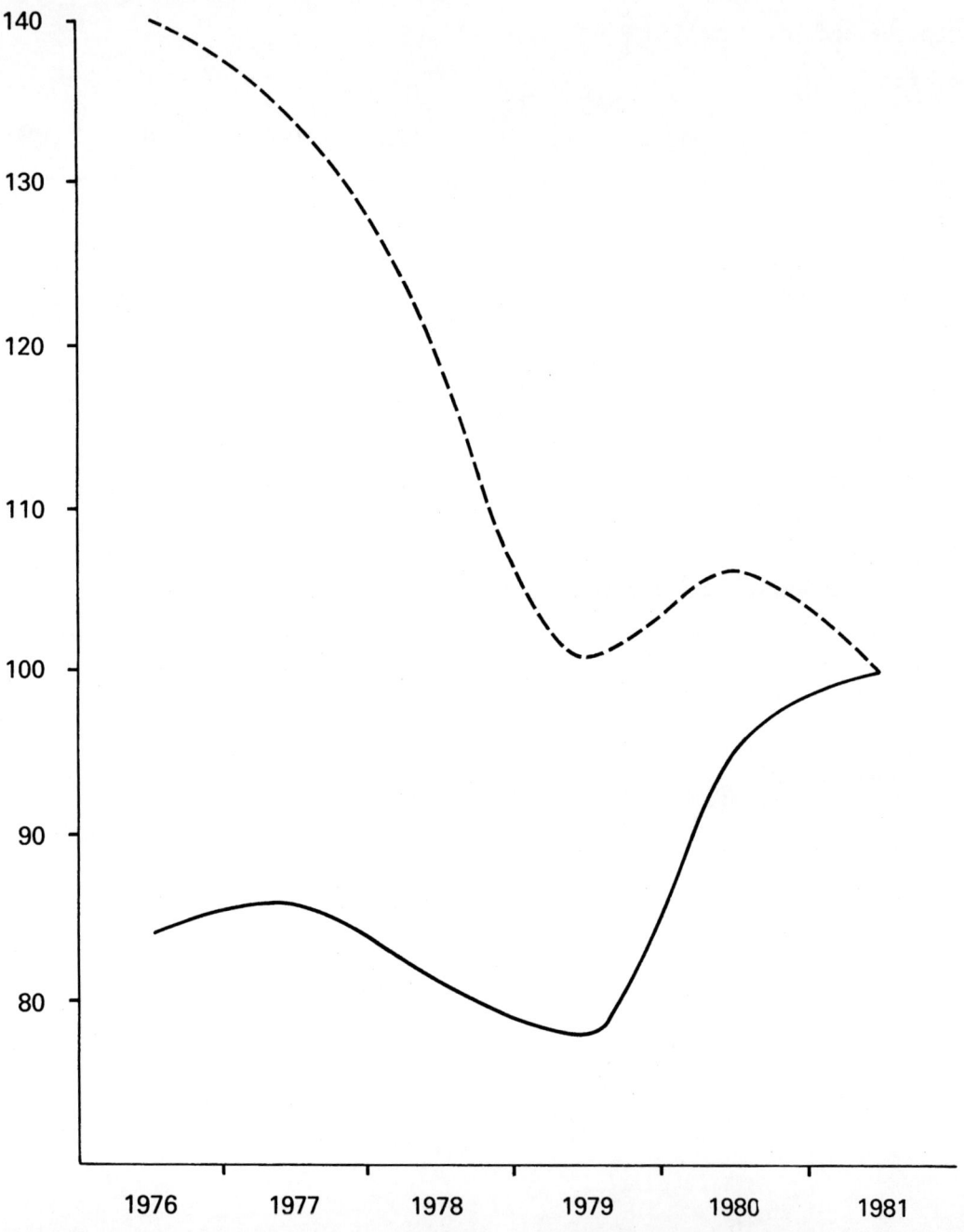

SUPPLY AND DEMAND BY MAIN MARKET AREA

	UK	EC (Ten)	Japan	USA
Production (1979/80 Averages) '000 tonnes				
Mine Production gross weight	-	15	82	-
contained mn	-	5	29	-
Ferro manganese	94.5	1016	625	230
of which: Low carbon	-	116	57	n/a
Medium carbon	) 94.5	900	95	n/a
High carbon	)		473	n/a
Metal				24.7
Net Imports (1979/80 Averages) '000 tonnes				
Manganese ore gross weight	335	2993	1721	543
Ferro manganese	66.4	274.8	10.8	647
Metal	3.2	13	n/a	6.1
Source of Net Imports (%)				
Manganese ore				
Australia	2	5	34	26
European Community	5	-	-	-
S Africa	55	48	42	27
Brazil	32	11	9	15
Gabon	1	29	8	25
Ghana	2	2		
Mexico		1	4	4
Morocco	2			3
Others	1	2	3	
Ferro Manganese				
Australia				3
Canada				2
European Community	17			28
Japan				3
Norway	40	56		6
Portugal		5		3
S Africa	38	27		41
Spain		7		2
Yugoslavia		1		2
Brazil				3
Mexico				6
Others		4		1

	UK	EC (Ten)	Japan	USA
Net Exports (1979/80 Averages) '000 tonnes				
Manganese ore	6	21	-	-
Ferro manganese	4	238.6	68	17
Metal	0.4	0.9	n/a	6.1
Consumption (1979/80 Averages) '000 tonnes				
Manganese ore	329 (iron & steel)	3051	1521.4	1108
Ferro manganese	137.5 (iron & steel)	1031.4	812.4	800.9
Import Dependence (manganese content)				
Imports as % of consumption	100	99.5	99	100
Imports as % of consumption and net exports	100	99.5	99	100
Share of World Consumption %				
Total world (approx.)				
Manganese ore	1	12	6	4
Ferro manganese	3	19	15	15
Consumption Growth % p.a.				
1970s				
Manganese ore	-2.7	-0.2	1.5	-6.0
Ferro manganese	-4.7	0.5	2.1	-1.6

****

MERCURY

WORLD RESERVES
('000 76 lb flasks and % of total)

Developed			Less Developed			Centrally Planned			Total
Canada	120	(2.7)	Algeria	350	(7.8)	China	500	(11.1)	
Italy	350	(7.8)	Mexico	250	(5.6)	USSR	500	(11.1)	
Spain	1450	(32.3)	Others	30	(0.7)	Others	10	(0.2)	
Turkey	60	(1.3)							
USA	350	(7.8)							
Yugoslavia	500	(11.1)							
Others	20	(0.4)							
Totals	2850	(63.5)		630	(14.0)		1010	(22.5)	4490

Identified world resources amount to 16.8 million flasks.

WORLD MINE PRODUCTION
('000 76 lb flasks and % of total 1979/80 Averages)

Developed			Less Developed			Centrally Planned			Total
Finland	1.3	(0.7)	Algeria	30	(15.7)	China	20	(10.5)	
W Germany	2.6	(1.4)	Dominican Rep.	0.5	(0.3)	Czecho-slovakia	4.8	(2.5)	
Spain	33.1	(17.4)	Mexico	1.7	(0.9)	USSR	61.5	(32.3)	
Turkey	4.9	(2.6)							
USA	30.1	(15.8)							
Totals	72.0	(37.8)		32.2	(16.9)		86.3	(45.3)	190.5

Note:- Italy's mine produced 22,278 flasks in 1976 but was closed by 1978. Yugoslavia's production (12,503 flasks in 1976) also ceased in 1977. Both mines closed in response to weak markets. The Italian mine re-opened on a modest scale in 1981, and plans are in hand for re-opening the Yugoslav mine.

RESERVE/PRODUCTION RATIOS

Static reserve life (years) : 24
Ratio of identified resources
to cumulative demand 1981-2000 4 : 1

CONSUMPTION

With increasingly tight environmental controls on mercury usage, demand declined considerably in the 1970s and a growing percentage of that demand was met from secondary recovery. Statistics on total European demand are not available.

	1979/80 Averages '000 flasks	% p.a. Growth rates 1970s
Japan	8.3	-11.5
United States	60.6	-1.4

END USE PATTERNS 1980 (USA) %

Electrical apparatus	51
Mildew proofing paint	14
Electrolytic production of chlorine/caustic soda	13
Industrial and control instruments	10
Others	7

VALUE OF CONTAINED METAL IN ANNUAL PRODUCTION

$79 million (at average 1981 prices)

SUBSTITUTES

Few satisfactory substitutes for applications in electrical apparatus and industrial and control instruments.

The diaphragm cell can replace cells using mercury in the chlor-alkali industry although it produces a lower quality caustic soda product.

Organic mildewicides are being substituted in latex paints; plastic paint and copper oxide paint are being used to protect ship hulls.

TECHNICAL POSSIBILITIES

Environmental considerations are likely to encourage conservation and recycling.

Design changes in mercury cell and improvements in diaphragm cell could modify consumption.

PRICES

	1976	1977	1978	1979	1980	1981
New York Dealer Price 99.99% $/flask of 76 lb 20 + flask lots	121.3	135.7	153.3	281.1	389.4	413.9

Until 1978/79, mainly dealer markets; producer pricing has become more important since then, particularly outside US.

MARKETING ARRANGEMENTS

Spain, Italy and Yugoslavia and Algeria belong to Mercury Producers Association, ASSIMER. In last four years ASSIMER has become increasingly important in contributing to stability and strength of market, particularly in the face of falling demand and, by restricting sales, prices have been maintained at a higher relative level to most metals.

Index Numbers 1981 = 100

The solid line gives prices in money terms and the dotted line gives prices in 'real' 1981 terms

MERCURY
New York Dealer, 99·99%

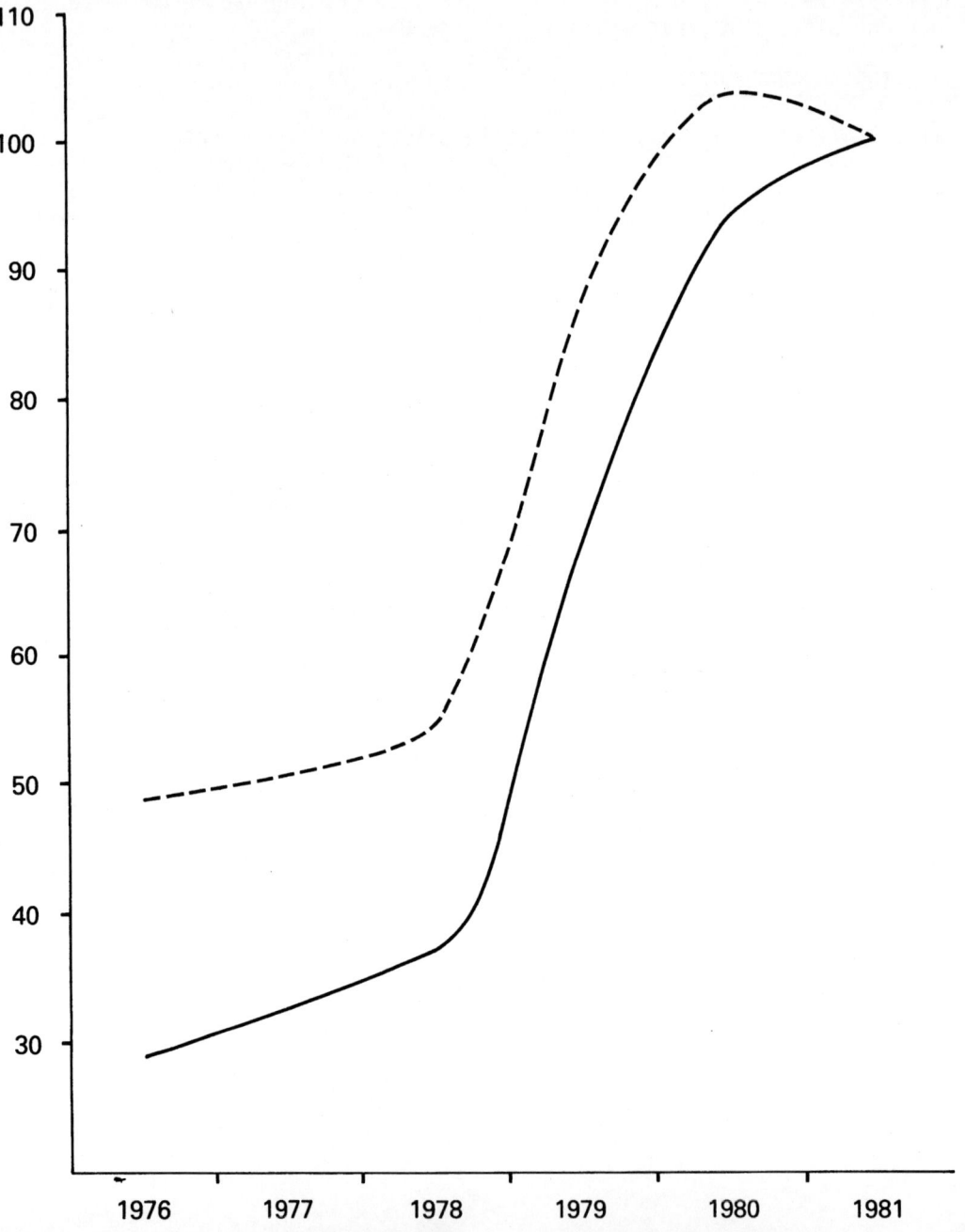

SUPPLY AND DEMAND BY MAIN MARKET AREA

	UK	EC (Ten)	Japan	USA
Production (1979/80 Averages) '000 76lb flasks				
Primary	-	2.57 (a)	-	30.08
Secondary	n/a	n/a	1.39	5.54
GSA releases	-	-	-	10.65
(a) excludes Italian sales from stock				
Net Imports (1979/80 Averages) '000 76lb flasks	7.8	26.77	2.9	17.93
Source of Net Imports (%)				
Canada		1		13
European Community	60			12
Finland		2		
Japan		2		33
Spain	26	46	12	33
Turkey		5		
USA	3	2		
China	3	19		
USSR		4		
Algeria	1	10	71	..
Mexico		1	17	4
Others	7	8		5
Net Exports (1979/80 Averages) '000 76lb flasks	3.14	9.51	12.5 (mainly secondary from stock)	n/a
Consumption (1979/80 Averages) '000 76lb flasks	c.5 (primary only)	c.20-30 (primary only)	8.3	60.59

	UK	EC (Ten)	Japan	USA
Import Dependence				
Imports as % of consumption	100 (primary only)	100 (primary only)	70	30
Imports as % of consumption and net exports	100 (primary only)	91 (primary only)	28	30

Share of World Consumption %

Because of the substantial tonnages of mercury consumed from secondary recovery or from stocks, and the lack of complete statistics thereon, reliable estimates of shares of world consumption cannot be made.

Consumption Growth % p.a.

	UK	EC (Ten)	Japan	USA
1970s	9.7 (primary only)	declined rapidly	-11.5	-1.4

****

MOLYBDENUM

WORLD RESERVES
('000 tonnes of metal and % of total)

Developed			Less Developed			Centrally Planned			Total
Canada	635	(6.4)	Chile	2450	(24.9)	China	225	(2.3)	
USA	5350	(54.3)	Iran	135	(1.4)	USSR	680	(6.9)	
			Mexico	135	(1.4)	Others	5	(0.1)	
			Peru	225	(2.3)				
			Others	5	(0.1)				
Totals	5985	(60.7)		2950	(30.0)		910	(9.3)	9845

Identified resources amount to almost 30 million tonnes.

WORLD MINE PRODUCTION
(tonnes of metal and % of total 1979/80 Averages)

Developed			Less Developed			Centrally Planned			Total
Canada	11686	(11.0)	Chile	13450	(12.7)	Bulgaria	150	(0.1)	
Japan	123	(0.1)	S Korea	226	(0.2)	China	2000	(1.9)	
USA	66826	(63.1)	Mexico	53	(0.1)	USSR	10300	(9.7)	
			Peru	1090	(1.0)				
			Philippines	132	(0.1)				
Totals	78635	(74.2)		14951	(14.1)		12450	(11.7)	106036

RESERVE/PRODUCTION RATIOS

Static reserve life (years)   : 93
Ratio of identified resources
to cumulative demand 1981-2000 : 8 : 1

CONSUMPTION

Molybdenum in all forms

	1979/80 Averages tonnes	% p.a. Growth rates 1970s
European Community	c.26500	2.3
Japan	12020	4.4
United States	30490	3.1
Other Countries	9690	2.4
Total Western world (excluding exports to Eastern countries)	78700	2.5

Source: Amax publications

END USE PATTERNS 1980 (USA) %

Machinery	34
Oil and Gas industry	20
Transport	17
Chemicals	13
Electrical	8
Others	8

VALUE OF CONTAINED METAL IN ANNUAL PRODUCTION

$1.6 billion (at average 1981 prices)

SUBSTITUTES

Potential substitutes in alloy steel include boron, chromium, manganese, columbium, vanadium and nickel. Tungsten can be used in tool steels and, along with tantalum, in certain refractory metal uses. Graphite can replace molybdenum for refractory elements in some electric furnaces. Chrome orange, cadmium red and organic orange pigments are substitutes for molybdenum orange. Most of the above alternatives to molybdenum suffer losses in efficiency. Heat treatment of alloy steels is an alternative to molybdenum.

TECHNICAL POSSIBILITIES

Increased Mo recovery through improvement in efficiency of flotation techniques.

Development and application of new Mo-containing steels and alloys particularly if resistance to oxidation at high temperatures improves solar energy panels.

PRICES

	1976	1977	1978	1979	1980	1981
Climax Concentrate 95% MoS$_2$ $/lb Mo	2.94	3.68	4.52	7.61	9.78	8.48
Climax Oxide molybdic oxide (producer price) 85.5% oxide min $/lb	3.25	3.7	4.9	6.1	8.9	9.0
Dealer Oxide molybdic trioxide, export $/lb (range)	2.08-4.2	4.1-6.2	5.6-18.0	14.25-32.25	6.7-16.5	3.3-8.4

Mainly producer priced, with dealer market which can influence producer
price movements. By-product material normally sold at discounts from Climax
price. Ferro molybdenum prices linked to concentrate price. Dealer market
gained in importance with 1978/79 shortages, and US export prices rose
faster than US domestic prices because of controls. Differential removed in
1980. Dealer price currently below producer in very depressed market.

MARKETING ARRANGEMENTS

Less than 10 mines in US, Canada and Chile account for a large percentage of
world's principal production with American Metal Climax having 40-50% of
total output. By-product production from copper more diffused. In boom
times trend towards increased by-product output to reduce copper production
costs. Recent low prices have discouraged new projects and brought
substantial production cutbacks in the US and Canada.

Index Numbers 1981 = 100

The solid line gives prices in money terms and the dotted line gives prices in 'real' 1981 terms

MOLYBDENUM
Climax Concentrate, 95% MoS$_2$

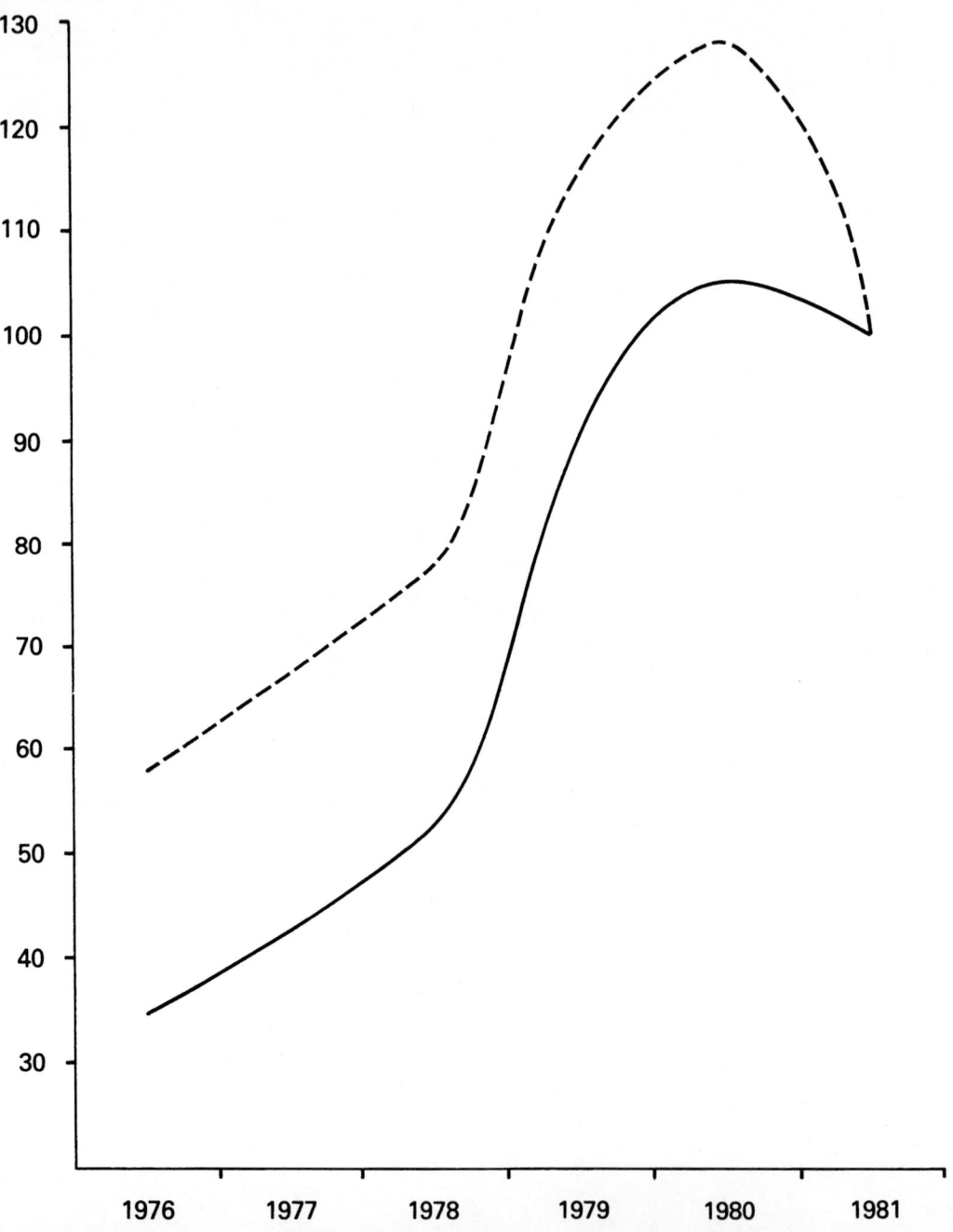

SUPPLY AND DEMAND IN MAIN MARKET AREAS

	UK	EC (Ten)	Japan	USA
Production (1979/80 Averages) tonnes				
Mine production (mo content)	-	-	123	66826
Ferro molybdenum (gross weight)	n/a	n/a	3765	n/a
Molybdic oxide inc. briquettes	n/a	n/a	5938	51207 (a)

(a) Includes use to make other products

Net Imports (1979/80 Averages) tonnes	UK	EC (Ten)	Japan	USA
Ores and concentrates (mo content)	7840	26615	20122	942
Ferro molybdenum	681	1067	497	17
Other molybdenum products (mo content)			720	1290
Oxides	290	426		
Metal	255	1248		

Source of Net Imports (%)

Ores and concentrates

	UK	EC (Ten)	Japan	USA
Canada	18	26	26	
European Community	36			
Sweden		2		
United States	15	44	53	
Chile	27	23	15	
Peru		3		
Others	4		6	

Net Exports (1979/80 Averages) tonnes	UK	EC (Ten)	Japan	USA
Ores and concentrates (mo content)	1780	3603	-	31856
Ferro molybdenum (gross weight)	2068	1874	-	1720
Metal (gross weight)	253	192.5	54	1007
Compounds (gross weight)	1159 (oxides)	1689 (inc. oxides)	-	4637

Consumption (1979/80 Averages)
  tonnes

  All forms (mo content)   5300   c.26500   12020   30490

Import Dependence

  Imports as % of consumption   100   100   99   -
  Imports as % of consumption and
                    net exports   100   100   99   -

Share of World Consumption %

  Western world   7   34   15   38

Consumption Growth % p.a.

  1970s   -3.5   2.3   4.4   3.1

****

NICKEL

WORLD RESERVES
('000 tonnes of metal and % of total)

Developed			Less Developed			Centrally Planned			Total
Australia	5080	(9.3)	Brazil	420	(0.8)	Cuba	3100	(5.7)	
Canada	7890	(14.5)	Colombia	820	(1.5)	USSR	7350	(13.5)	
S Africa	1500	(2.8)	Dominican Rep.	1000	(1.8)				
USA	180	(0.3)	Guatemala	270	(0.5)				
Others	200	(0.4)	Indonesia	7080	(13.0)				
			New Caledonia	13610	(25.0)				
			Philippines	5170	(9.5)				
			Others	830	(1.5)				
Totals	14850	(27.2)		29200	(53.6)		10450	(19.2)	54500

Identified world resources of nickel in deposits averaging 1% nickel or more exceed 130 million tonnes of which 80% is in laterites. Resources of lower grade deposits are very large, and there are extensive sea bed resources of nickel in manganese nodules.

WORLD MINE PRODUCTION
('000 tonnes of contained metal and % of total 1979/80 Averages)

Developed			Less Developed			Centrally Planned			Total
Australia	69.9	(9.8)	Botswana	15.8	(2.2)	China	11.0	(1.5)	
Canada	160.7(a)	(22.5)	Brazil	2.7	(0.4)	Cuba	35.3	(4.9)	
Finland	6.1	(0.9)	Burma	0.1	(..)	USSR	144.0	(20.2)	
Greece	14.3	(2.0)	Dominican Rep.	20.3	(2.8)	Others	12.4	(1.7)	
Norway	0.6	(0.1)	Guatemala	6.6	(0.9)				
S Africa	25.5	(3.6)	Indonesia	38.1	(5.3)				
USA	12.5	(1.8)	Morocco	0.5	(0.1)				
Yugoslavia	1.5	(0.2)	New Caledonia	84.6	(11.9)				
			Philippines	34.1	(4.8)				
			Zimbabwe	16.8	(2.4)				
Totals	291.0	(40.8)		219.6	(30.8)		202.7	(28.4)	713.3

(a) Canada's 1979 output was greatly reduced by strikes. The 1975/77 average was 238.5. Its share of 1980 world output was just over 26.

WORLD REFINED METAL PRODUCTION
('000 tonnes of metal and % of total 1979/80 Averages)

Developed			Less Developed			Centrally Planned			Total
Australia	38.9	(5.4)	Brazil	2.3	(0.3)	China	11.0	(1.5)	
Canada	118.0(a)	(16.5)	Dominican Rep.	20.9	(2.9)	Cuba	19.6	(2.7)	
France	6.2	(0.9)	Indonesia	4.2	(0.6)	USSR	169.0	(23.6)	
Finland	12.1	(1.7)	New Caledonia	31.5	(4.4)	Others	13.4	(1.9)	
Greece	14.3	(2.0)	Philippines	20.7	(2.9)				
Japan	107.6	(15.0)	Zimbabwe	14.8	(2.1)				
Norway	33.8	(4.7)							
S Africa	18.0	(2.5)							
UK	19.1	(2.7)							
USA	40.1	(5.6)							
Totals	408.1	(57.0)		94.4	(13.2)		213.0	(29.8)	715.5

(a) Canada's 1979 output was greatly reduced by strikes. It was 152.3
in 1980 (20% of the world total).

RESERVE/PRODUCTION RATIO

Static reserve life (years)   : 76
Ratio of identified resources
to cumulative demand 1981-2000   : $5\frac{1}{2}$: 1 (land based only)

CONSUMPTION

	1979/80 Averages '000 tonnes	% p.a. Growth rates 1960s	1970s
European Community (Ten)	172.4	5.9	3.4
Japan	127.0	18.9	4.3
United States	163.2	3.2	1.8
Others	99.7	11.0	5.7
Total Western world	562.3	7.0	3.4
Total world	751.1	6.9	3.6

END USE PATTERNS 1980 (USA) %

Stainless and alloy steel	45	Transport	25
Nonferrous alloys	30	Chemical Industry	15
Electroplating	15	Electrical equipment	15
Others	10	Construction	10
		Fabricated metal products	10
		Other	25

VALUE OF CONTAINED METAL IN ANNUAL PRODUCTION

$47 billion (refined metal at 1981 average prices)

SUBSTITUTES

The use of alternative materials tends to be more expensive or requires sacrifice in chemical or physical characteristics, and hence performance. However alternative materials are available to replace nickel in all its uses. Alloy substitutes are normally other 'steel' industry metals such as molybdenum, columbium and manganese. Platinum, cobalt and copper can be used in some catalysts. Titanium and many plastics can compete for markets where corrosion-resistance is important.

TECHNICAL POSSIBILITIES

Deep sea nodules.

Development of new nickel-bearing alloys.

PRICES

	1976	1977	1978	1979	1980	1981
Cathode US $/lb						
US Producer	2.25	2.36	2.09	2.71	3.42	3.46
US Dealer	2.08	2.08	1.93	2.58	3.08	2.87
LME Cash £/tonne	-	-	-	2698.82*	2805.51	2948.47

* Six months only

Producer pricing with dealer market. Discounting in weak demand periods. Breakeven costs influenced by associated by-product revenues. Triannual labour agreement in Canadian industry has short term effect on price. London Metal Exchange quotation introduced in mid-1979.

MARKETING ARRANGEMENTS

International Nickel (Inco) retains over one-third of Western World market, with Falconbridge, Imetal, Western Mining as other major producers. Inco normally market leader with strong influence on market prices. Other producers affected by higher and faster rising costs. Lateritic mines adversely affected by energy cost increases. Production cutbacks in major producing areas currently in operation. Influence of major producers has weakened in recent years.

Index Numbers 1981 = 100

The solid line gives prices in money terms and the dotted line gives prices in 'real' 1981 terms

NICKEL
US Producer, Cathode

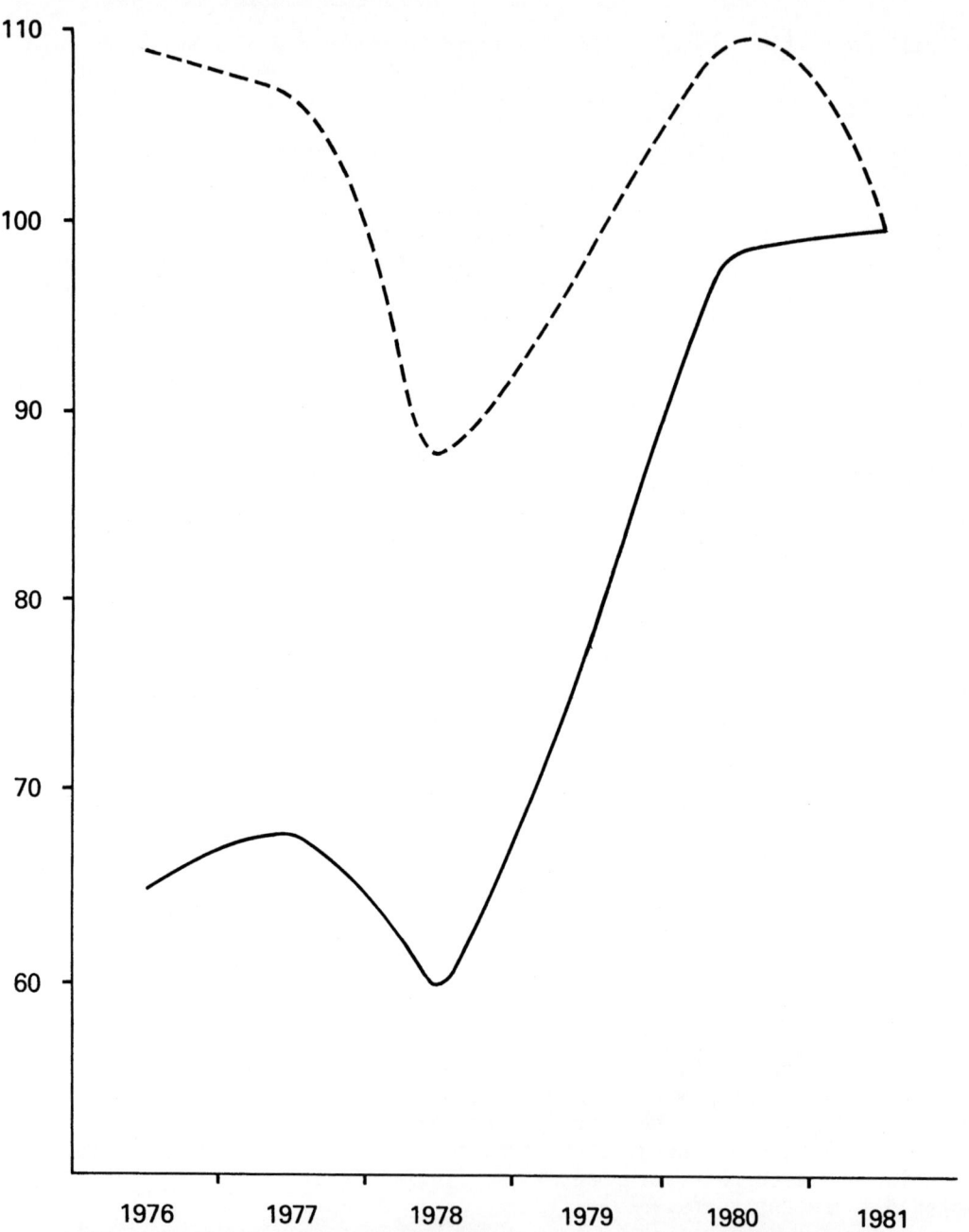

SUPPLY AND DEMAND BY MAIN MARKET AREA

	UK	EC (Ten)	Japan	USA
Production (1979/80 Averages) '000 tonnes ni content				
Mine production	-	14.3	-	12.5
Production of matte	-	-	-	-
Smelter Refinery/Production	19.1	40.05	107.6	40.15
of which Metal	19.1	25.75	24.9	28.6
Ferro and nickel oxide sinter	-	14.3	82.7	11.55
Net Imports (1979/80 Averages) '000 tonnes ni content				
Ores and concentrates	-	0.25	68.85	-
Matte and mixed sulphides	24.75	32.6	35.5	27.65
Ferro and nickel oxide sinter and metal	18.55	134.2	36.6	130.05
Source of Net Imports %				
All forms				
Australia	9	11	16	7
Canada	60	23	3	42
European Community	6	-	..	..
Finland	2	2	..	2
Japan		1		1
Norway	7	6	2	10
S Africa	1	8	1	6
United States	..	7	1	
Eastern Countries (inc. Cuba)	1	14	3	4
Botswana	-	-	-	10
Dominican Republic	1	5	3	5
Guatemala	12	3	1	
Indonesia		..	27	
New Caledonia		14	31	5
Philippines		4	11	7
Zimbabwe	..	1	..	..
Others	..	1	..	..
Net Exports (1979/80 Averages) '000 tonnes ni content				
Ores and concentrates	-	-	-	0.5
Matte and mixed sulphides	1.25	2.3	-	-
Ferro and nickel oxide sinter and metal	11.9	18.05	4.55	16.35

	UK	EC (Ten)	Japan	USA
Consumption (1979/80 Averages) '000 tonnes ni content				
All forms	28.9	172.4	127.0	163.2
Import Dependence				
Imports as % of consumption	100	97	100	97
Imports as % of consumption and net exports	100	87	100	88
Share of World Consumption %				
Western world	5	31	23	29
Total world	4	23	17	22
Consumption Growth % p.a.				
1960s	1.6	5.9	18.9	3.2
1970s	-0.3	3.4	4.3	1.8

****

NIOBIUM

WORLD RESERVES
('000 tonnes of metal and % of total)

Developed			Less Developed			Centrally Planned			Total
Canada	120	(2.6)	Brazil	3220	(69.9)	USSR	680	(14.8)	
Europe	90	(2.0)	Nigeria	65	(1.4)				
			Zaire	30	(0.7)				
			Others	400	(8.7)				
Totals	210	(4.6)		3715	(80.7)		680	(14.8)	4605

World resources are estimated at some 17,250 million tonnes.

WORLD MINE PRODUCTION
(tonnes of contained metal and % of total 1979/80 Averages)

Developed			Less Developed			Centrally Planned		Total
Australia	36	(0.2)	Brazil	12415	(85.5)	China	n/a	
Canada	1700	(11.7)	Malaysia	10	(0.1)	USSR	n/a	
Portugal	1	(..)	Mozambique	6	(..)			
Spain	n/a	n/a	Nigeria	247	(1.7)			
			Rwanda	16	(0.1)			
			Thailand	74	(0.5)			
			Uganda	1	(..)			
			Zaire	12	(0.1)			
			Zambia	n/a	n/a			
			Zimbabwe	3	(..)			
Totals	1737	(12.0)		12784	(88.0)		n/a	14521 (W world)

Note:- Reliable estimates of production in centrally planned economies are not available. The total and shares are for the Western world only.

RESERVE/PRODUCTION RATIOS

Static reserve life (years)   329
Ratio of identified resources
to cumulative demand 1981-2000   38 : 1

CONSUMPTION

Reliable statistics are not available for most areas but broad orders of magnitude are as follows for contained niobium in all forms.

	1979/80 Averages tonnes	% p.a. Growth rates 1970s
European Community	3000	approx. 5 to 8
Japan	1800-2000	12.1
United States	3325	4.1

END USE PATTERNS 1980 (USA) %

Transport	32
Construction	31
Oil and Gas industries	16
Machinery	11
Other	10

VALUE OF CONTAINED METAL IN ANNUAL PRODUCTION

$100 million (Western world only at average 1981 prices)

SUBSTITUTES

Vanadium, titanium and molybdenum in HSLA steels. Tantalum competes, though at higher cost, in superalloys. Titanium can be used in stainless steel.

Substitutes usually lower performance and/or cost effectiveness.

TECHNICAL POSSIBILITIES

Refinements in beneficiating and processing techniques are giving products of higher purity or different composition.

Development of new steels, superalloys and super-conductors.

Catalytic applications.

PRICES

	1976	1977	1978	1979	1980	1981
Ore: Pyrochlore ¢/lb contained Cb_2O_5	n/a	255.0 (from May)	255.0	255.0	255.0	325

1977-1980 inclusive Brazilian pyrochlore 50-55% Cb_2O_5

    1981   Canadian pyrochlore 57-62% Cb_2O_5

Mainly producer price basis, and nominal price changes infrequent. Concentrate producers have low costs relative to prices. Columbite output (as tin by-product) dependent on tin production.

MARKETING ARRANGEMENTS

Brazilian Araxa mine (Companhia Brasileria de Metalurgica e Mineracao) and Niobec in Canada the major concentrate producers, dominate the market. Substantial forward integration into ferrocolumbium.

Metal producers separate.

Index Numbers 1981 = 100

The solid line gives prices in money terms and the dotted line gives prices in 'real' 1981 terms

NIOBIUM
Pyrochlore

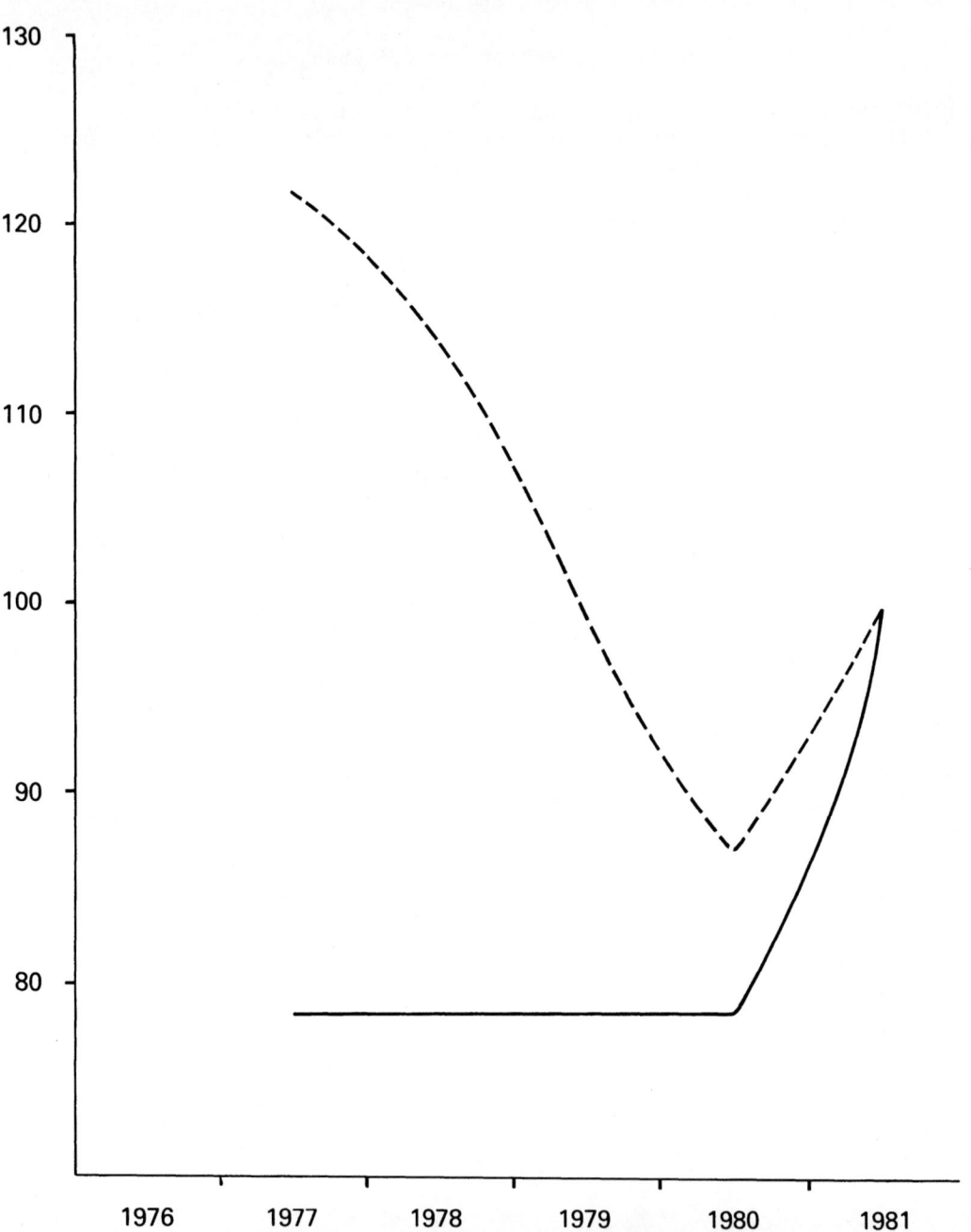

SUPPLY AND DEMAND BY MAIN MARKET AREA

	UK	EC (Ten)	Japan	USA
Production (1979/80 Averages) tonnes				
Mine production	-	-	-	-
Ferro niobium (approx. gross weight)	n/a	n/a	1248	[1043]
Ferro niobium (approx. nb content)	n/a	n/a	811	678
Net Imports (1979/80 Averages) tonnes				
Ores and concentrates (gross weight)	376	4208	1712	909 (nb cont.)
Ferro niobium (gross weight)	794	(4800) (approx.)	1659	2593 (nb cont.)
Tin slags - nb content	n/a	n/a	n/a	c.500
Metal and alloys	28.5	32.5	n/a	17

Source of Net Imports %

Ores and concentrates

	UK	EC (Ten)	Japan	USA
Australia		1		
Canada	4	41	53	31
European Community	2	-		4
S Africa		2		
United States		17	1	
China			1	
Brazil	91	28	38	29
Malaysia		1		3
Nigeria	2	1	7	23
Thailand		2		1

Ferro niobium

	UK	EC (Ten)	Japan	USA
Austria	9		-	-
European Community	8	-	7	-
Brazil	83	nearly 100	93	100

Net Exports (1979/80 Averages)
 tonnes

	UK	EC (Ten)	Japan	USA
Ores and concentrates (gross weight)	75	1567	-	-
Ferro niobium (gross weight)	418	n/a	n/a	-
Metal compounds and alloys	-	0.5	n/a	50

	UK	EC (Ten)	Japan	USA
Consumption (1979/80 Averages) tonnes				
Niobium metal	n/a	n/a	n/a	n/a
Ferro niobium (nb content)	430 (iron & steel usage)	n/a	1542 (gross 2373)	2912
Total all forms	c.450	c.3000	c.1800	3323
Import Dependence				
Imports as % of consumption	100	100	100	100
Imports as % of consumption and net exports	100	100	100	100
Share of World Consumption %				
Western world (approx.)	4	30	18	33
Consumption Growth % p.a.				
1970s	-5	approx. 5 to 8 (ferro only)	12.1	4.1

****

PHOSPHATE

WORLD RESERVES
(million tonnes and % of total)

Developed			Less Developed			Centrally Planned			Total
Australia	2770	(3.4)	Algeria	1000	(1.2)	China 10000	(12.3)		
Finland	565	(0.7)	Brazil	800	(1.0)	N Korea 90	(7.1)		
S Africa	700	(0.8)	Christmas Is.	62	(0.1)	USSR 6300	(7.7)		
Turkey	150	(0.2)	Colombia	56	(0.1)	Vietnam 100	(0.1)		
USA	8000	(9.8)	Egypt	3000	(3.7)				
			India	100	(0.1)				
			Israel	150	(0.2)				
			Jordan	1100	(1.3)				
			Mexico	1034	(1.3)				
			Morocco	44000	(53.9)				
			Nauru	20	(..)				
			Senegal	75	(0.1)				
			Syria	833	(1.0)				
			Togo	110	(0.1)				
			Tunisia	500	(0.6)				
			Uganda	130	(0.2)				
			Zimbabwe	50	(..)				
Totals	12185	(14.9)		53020	(64.9)		16490	(20.2)	81695

Identified world resources total almost 130 million tonnes.

WORLD MINE PRODUCTION
(million tonnes and % of total 1979/80 Averages)

Developed			Less Developed			Centrally Planned			Total
S Africa	3.25	(2.4)	Algeria	1.05	(0.8)	China 6.85	(5.1)		
USA	53.01	(39.5)	Brazil	2.26	(1.7)	N Korea 0.50	(0.4)		
Others	0.17	(0.1)	Christmas Is.	1.54	(1.1)	USSR 25.84	(19.3)		
			Egypt	0.62	(0.5)	Vietnam 0.40	(0.3)		
			India	0.55	(0.4)				
			Israel	2.41	(1.8)				
			Jordan	3.53	(2.6)				
			Mexico	0.25	(0.2)				
			Morocco	19.43	(14.5)				
			Nauru	1.95	(1.5)				
			Senegal	1.67	(1.2)				
			Syria	1.30	(1.0)				
			Togo	2.93	(2.2)				
			Tunisia	4.37	(3.3)				
			Zimbabwe	0.14	(0.1)				
			Others	0.12	(0.1)				
Totals	56.43	(42.1)		44.12	(32.9)		33.59	(25.0)	134.14

RESERVE/PRODUCTION RATIOS

Static reserve life (years)   : 609
Ratio of identified resources
to cumulative demand 1981-2000   : greater than 10 : 1

CONSUMPTION

	1979/80 Averages '000 tonnes	% p.a. Growth rates 1970s
European Community	c.17250	1.7
Japan	c.2400	0.8
United States	40191	4.3

END USE PATTERNS 1980 (USA) %

Fertilisers and animal feed supplements	88
Industrial and food grade products	12

VALUE OF ANNUAL PRODUCTION

$6 billion (at average 1981 prices)

SUBSTITUTES

No substitutes for agricultural applications.

TECHNICAL POSSIBILITIES

Reduction in level of sodium tripolyphosphate in detergents by substitution with other compounds.

PRICES

	1976	1977	1978	1979	1980	1981
Florida, land pebble, export 74-75% BPL, $/st	44.5	44.5	39.0	25.5	31.3	37.0
Moroccan 75-77% BPL fas Casablanca $/tonne	48.5	48.5	48.5	48.5	48.5	48.5

NB: Moroccan price is nominal only

Prices fixed on contract basis depending on quality and grade. Phosphate fertiliser contracts usually short term in duration, acid business has annual contracts with six months' pricing. Prices not published, above only guidelines. US prices usually lag behind Moroccan.

MARKETING ARRANGEMENTS

Fertiliser and acid markets now supplied mainly by large integrated producers with captive phosphate rock. Morocco, USSR and USA produce over 70% of world production but new developments, including attendant acid and fertiliser plants, coming onstream worldwide and are diversifying supply sources.

Index Numbers 1981 = 100

The solid line gives prices in money terms and the dotted line gives prices in 'real' 1981 terms

PHOSPHATE ROCK
Floride, export, 74–75% BPL

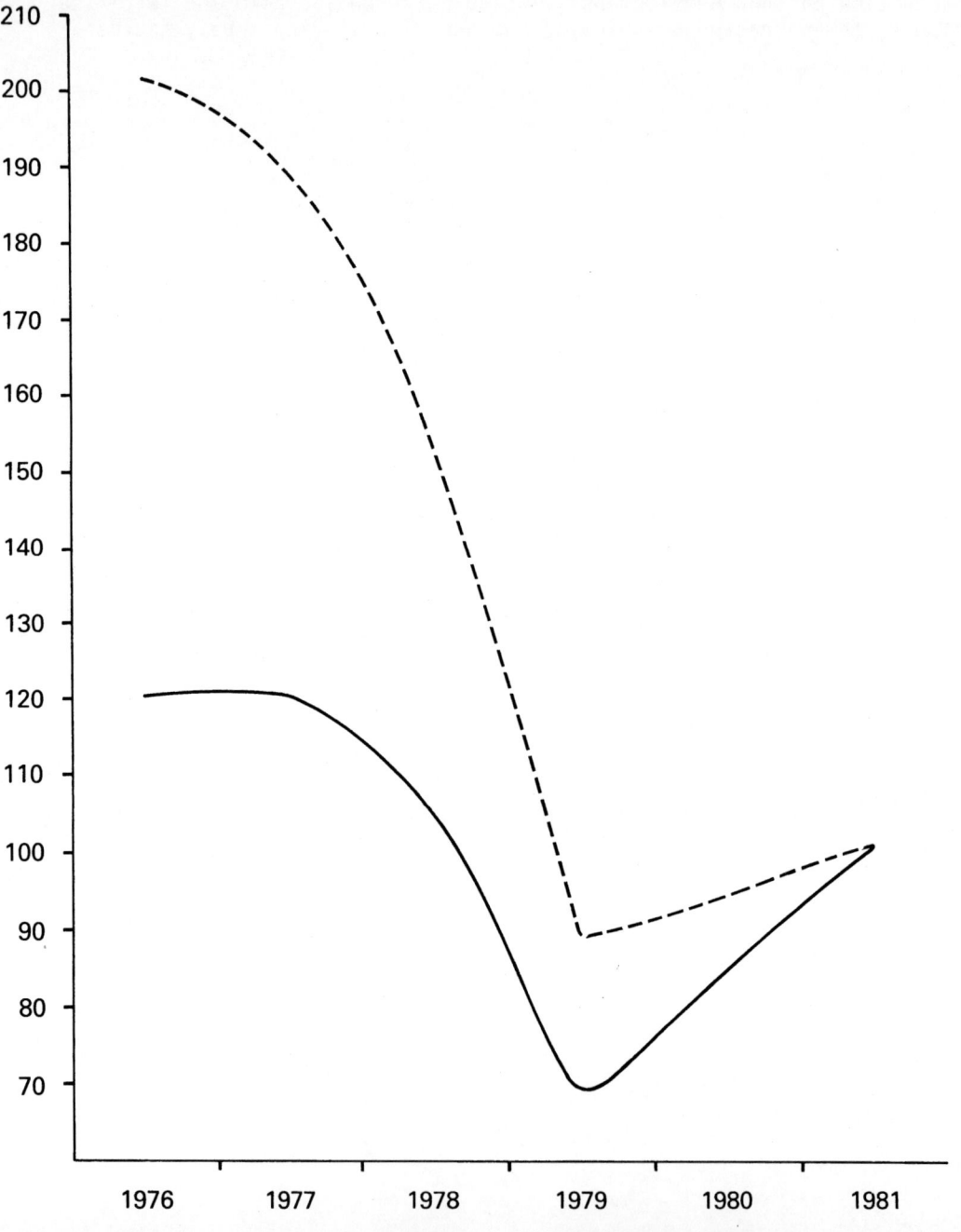

SUPPLY AND DEMAND BY MAIN MARKET AREA

	UK	EC (Ten)	Japan	USA
Production (1979/80 Averages) '000 tonnes				
Mine production	-	10.5	-	53013 (marketable)
Net Imports (1979/80 Averages) '000 tonnes				
Phosphate rock (gross)	1753	17200	2795	686
Superphosphates	59	583	27	-
Basic slag	15	10	-	-
Source of Net Imports %				
Phosphate rock (gross)				
European Community	1			-
United States		24	59	
USSR		2		
Algeria		1		
Israel		7		
Jordan		1	9	
Morocco	47	39	24	
Nauru		-	4	
Senegal	23	5	3	
Togo		11		
Tunisia		3		
Others and unidentified	29	7	1	
Net Exports (1979/80 Averages) '000 tonnes				
Phosphate rock	..	3	-	14317
Superphosphates	32.5	85	-	-
Basic slag	1.1	234	-	-
Consumption (1979/80 Averages) '000 tonnes				
Phosphate rock	c.1750	c.17250	766.4 P_2O_5 cont. = c.2400 gross	40191

	UK	EC (Ten)	Japan	USA
Import Dependence				
Imports as % of consumption	100	99	100	2
Imports as % of consumption and net exports	100	100	100	1
Share of World Consumption %				
Total world (approx.)	1.5	13	2	30
Consumption Growth % p.a.				
1970s	0.6 (based on rock imports)	1.7 (based on rock imports)	0.8	4.3

****

PLATINUM GROUP
(Platinum, Palladium, Iridium, Osmium, Rhodium, Ruthenium)

WORLD RESERVES
(tonnes and % of total)

Developed			Less Developed		Centrally Planned		Total
Canada	280	(0.8)	Colombia	very small	USSR	6220 (16.7)	
S Africa	30170	(81.2)	Zimbabwe	see note			
USA	500	(1.3)					
Totals	30950 (83.3)			..		6220 (16.7)	37170

The different deposits of platinum group metals have markedly different ratios between the constituent metals. The US Bureau of Mines gives the following breakdowns for the main deposits (in percentage by weight).

		Platinum	Palladium	Iridium	Rhodium	Ruthenium	Osmium
Colombia		93	1	3	3	-	1
Canada	- Sudbury	43	45	2	4	4	2
S Africa	- Merensky	61	26	1	3	8	1
	Bushveld	47	32	2	7	11	1
USSR	- Norilsk	25	67	2	3	2	1
USA	- Stillwater	20	78	..	1	..	-
	- Duluth	18	78	1	2	1	-

The approximate breakdown of reserves between the different metals in the two main deposits is as follows (in tonnes).

	S Africa	USSR
Platinum	18205	1865
Palladium	7805	3730
Rhodium	1040	125
Others	3120	500
Totals	30170	6220

World resources of platinum group metals are around 100,000 tonnes. Zimbabwe's deposits are still regarded as resources rather than reserves.

WORLD MINE PRODUCTION
(Kilograms and % of total 1979/80 Averages)

Developed			Less Developed			Centrally Planned		Total
Australia	305	(0.1)	Colombia	424	(0.2)	USSR	100310 (48.4)	
Canada	9466	(4.6)	Ethiopia	3	..	China	n/a	
Finland	20	..	Indonesia	n/a				
Japan (a)	1181	(0.6)	Papua					
			New Guinea	n/a				
S Africa	95130	(45.9)	Philippines	n/a				
USA	166	(0.1)						
Yugoslavia	182	(0.1)						
Totals	106450	(51.4)		425	(0.2)		100310 (48.4)	207185

(a) Japanese smelter/refinery recovery from ores originating elsewhere (including Australia, Canada, Indonesia, Papua New Guinea and Philippines).

The estimated breakdown of 1979's production was, in percentages:-

Platinum	43
Palladium	47
Iridium	2
Ruthenium	5
Rhodium	3

RESERVE/PRODUCTION RATIOS

Static reserve life (years)   : 180
Ratio of identified resources
to cumulative demand 1981-2000   18 : 1

CONSUMPTION

This includes substantial secondary recovery.

	1979/80 Averages Kilograms	% p.a. Growth rates 1970s
Japan	60550	10.1
United States	91040	6.3

Statistics are not available for the European Community.

END USE PATTERNS 1980 (USA) %

Automotive	30
Electrical	26
Chemical	14
Rental	12
Other	18

VALUE OF CONTAINED METAL IN ANNUAL PRODUCTION

$1.9 billion (at average 1981 producer prices for the various metals, weighted accordingly to 1979 production split).

SUBSTITUTES

Usually easier to substitute metals of the platinum group for one another, especially in alloys than to use alternate materials.

Substitutes in electrical uses include tungsten, nickel, silver, gold and silicon carbide.

Alternative catalysts include nickel, molybdenum, tungsten, chromium, cobalt, vanadium, silver and rare-earth materials, but normally with efficiency and cost penalties. However rhenium has been used most satisfactorily for part of platinum in petroleum-refining catalysts.

Stainless steel and ceramics can be used where corrosion resistance is of primary concern.

TECHNICAL POSSIBILITIES

Recovery from radio-active waste.

Creation of artificial platinum group metals in nuclear power reactors.

New or improved engines and fuels, and electric cars, could reduce application in this field.

PRICES

	1976	1977	1978	1979	1980	1981

$/troy oz

Platinum

	1976	1977	1978	1979	1980	1981
US Producer	161.7	162.5	237.3	351.6	439.4	475.0
US Dealer	153.3	157.7	260.8	444.6	677.3	446.0
Palladium: US Producer	50.93	59.7	70.87	113.14	213.98	129.5
Iridium : US Producer	330	305	305	284.6	502.38	600.0
Osmium : US Producer	152.5	152.5	152.5	152.5	152.5	152.5
Rhodium : US Producer	352.65	446.02	509.72	739.4	766.67	641.67
Ruthenium: US Producer	62.5	62.5	62.5	45	45	45

Combination of producer and dealer pricing, with futures trading in the USA.
Can be subject to speculative activity.

MARKETING ARRANGEMENTS

USSR, S Africa (Rustenburg and Impala), and Canada (Inco) to a lesser
extent, control market. Integrated producers. Most mining is in
association with nickel-copper ores. S Africa controls producer price of
platinum and USSR that of palladium, and can influence world spot price by
curtailing production and purchasing excess metal. Market for platinum
group metals currently very depressed.

Index Numbers 1981 = 100

The solid line gives prices in money terms and the dotted line gives prices in 'real' 1981 terms

PLATINUM
US Producer

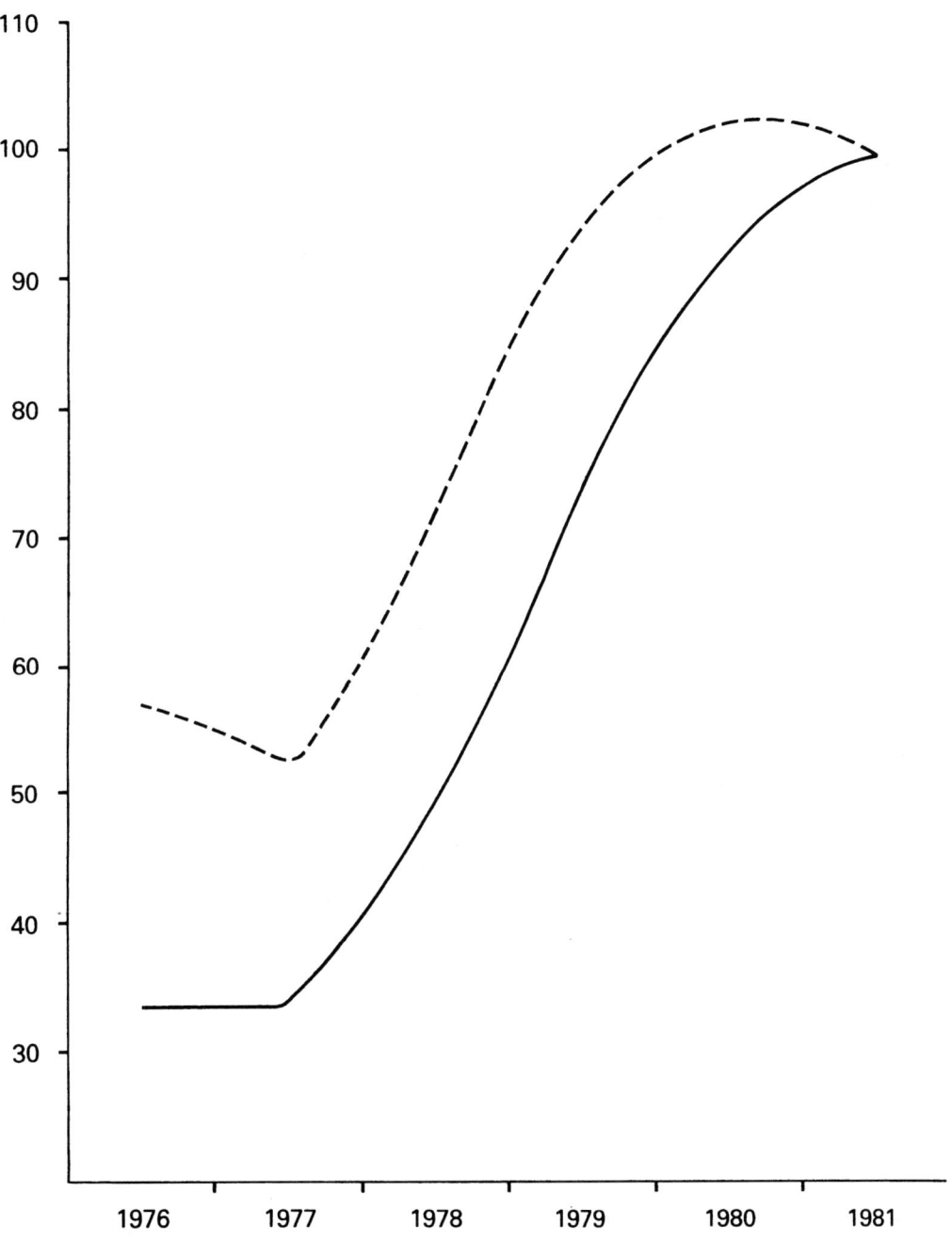

SUPPLY AND DEMAND BY MAIN MARKET AREA

	UK	EC (Ten)	Japan	USA

Production (1979/80 Averages)
Kilograms

	UK	EC (Ten)	Japan	USA
Mine production	-	-	-	166
Refinery production : New metal	n/a	n/a	1181	372
Secondary metal	n/a	n/a	n/a	9952

Net Imports (1979/80 Averages)
Kilograms of pt group

	UK	EC (Ten)	Japan	USA
Ores and concentrates	(a)	n/a)		
Waste and scrap	11050	n/a)	58700	108565
Unwrought metal inc. alloys	(15850	31500)		
Semi manufacturers inc. alloys	(	)		

(a) Value £90.3 million in 1979, and £188.7 million in 1980

Source of Net Imports (%)

	(unwrought & semi manufacturers & alloys)		(all forms)	(all forms)
Canada			-	3
European Community	38		9	16
			(UK only)	
Japan		2		1
S Africa	22	24	33	57
Switzerland		24	-	1
United States	31	21	10	-
USSR	3	29	44	15
Argentina		-	-	2
Others	6	-	4	5

Most ores and concentrates imported into the UK for refining are from South Africa, whose importance is thus much greater in world trade than this table suggests.

Net Exports (1979/80 Averages)
Kilograms

	UK	EC (Ten)	Japan	USA
Unwrought and semi manufactures inc. alloys ores and concentrates	48900	61500	1950	25887

	UK	EC (Ten)	Japan	USA
Consumption (1979/80 Averages)				
Kilograms	n/a	n/a	60550	91040
Import Dependence				
Imports as % of consumption	100 (exc. secondary)	100 (exc. secondary)	100 (exc. secondary)	100
Imports as % of consumption and net exports	100 (exc. secondary)	100 (exc. secondary)	100 (exc. secondary)	93

Share of World Consumption %

Insufficient information is published to complete this section, especially bearing in mind the large secondary recovery of platinum group metals.

Consumption Growth % p.a.

	UK	EC (Ten)	Japan	USA
1970s	n/a	n/a	10.1	6.3

****

POTASH

WORLD RESERVES

(million tonnes K_2O and % of total)

Developed			Less Developed			Centrally Planned			Total
Canada	14000	(68.7)	Brazil	50	(0.2)	China	100	(0.5)	
France	50	(0.2)	Chile	10	..	E Germany	800	(3.9)	
W Germany	500	(2.5)	Congo	20	(0.1)	USSR	4000	(19.6)	
Italy	10	..	Israel/Jordan	300	(1.5)				
Spain	60	(0.3)	Laos	20	(0.1)				
UK	60	(0.3)	Thailand	100	(0.5)				
USA	300	(1.5)							
Totals	14980	(73.5)		500	(2.5)		4900	(24.0)	20380

Estimated total world resources exceed 140,000 million tonnes, with two-thirds located in Saskatchewan. Much of these resources are only recoverable with solution mining techniques because of their depth.

WORLD PRODUCTION

(million tonnes of K_2O and % of total 1979/80 Averages)

Developed			Less Developed			Centrally Planned			Total
Canada	7.31	(27.2)	Chile	0.02	(0.1)	China	0.01	(..)	
France	1.93	(7.2)	Israel	0.76	(2.8)	E Germany	3.41	(12.7)	
W Germany	2.68	(10.0)				USSR	7.32	(27.2)	
Italy	0.18	(0.7)							
Spain	0.78	(2.9)							
UK	0.27	(1.0)							
USA	2.23	(8.3)							
Totals	15.38	(57.2)		0.78	(2.9)		10.74	(39.9)	26.90

RESERVE/PRODUCTION RATIOS

Static reserve life (years)   758
Ratio of identified resources
to cumulative demand 1981-2000 greater than 170 : 1

CONSUMPTION

	1979/80 Averages '000 tonnes K_2O	% p.a. Growth rates 1970s
European Community	5000 (approx.)	1.5 (approx.)
Japan	736 (1979 only)	1.0
United States	6672	4.5

END USE PATTERNS 1980 (USA) %

Fertiliser industry	95
Other (primarily caustic potash-chlorine plants)	5

VALUE OF ANNUAL PRODUCTION

$3.4 billion (at average 1981 prices)

SUBSTITUTES

No substitution for agricultural use.

TECHNICAL POSSIBILITIES

Recovery from low grade resources.

PRICES

	1976	1977	1978	1979	1980	1981
Muriate 62% K_2O, fob						
Saskatchewan Standard $/st	67.9	69.6	72.3	88.8	102.2	113.9

Producer list pricing for long term contracts. Discounting prevalent.

MARKETING ARRANGEMENTS

USSR and E Germany provide 40% of world supply, N America and W Europe 56%. USSR production fluctuates considerably with consequent effect on world market. High proportion of Canadian production, second after USSR, controlled by provincial Government (Saskatchewan).

Index Numbers 1981 = 100

The solid line gives prices in money terms and the dotted line gives prices in 'real' 1981 terms

POTASH
62% K₂O, Standard

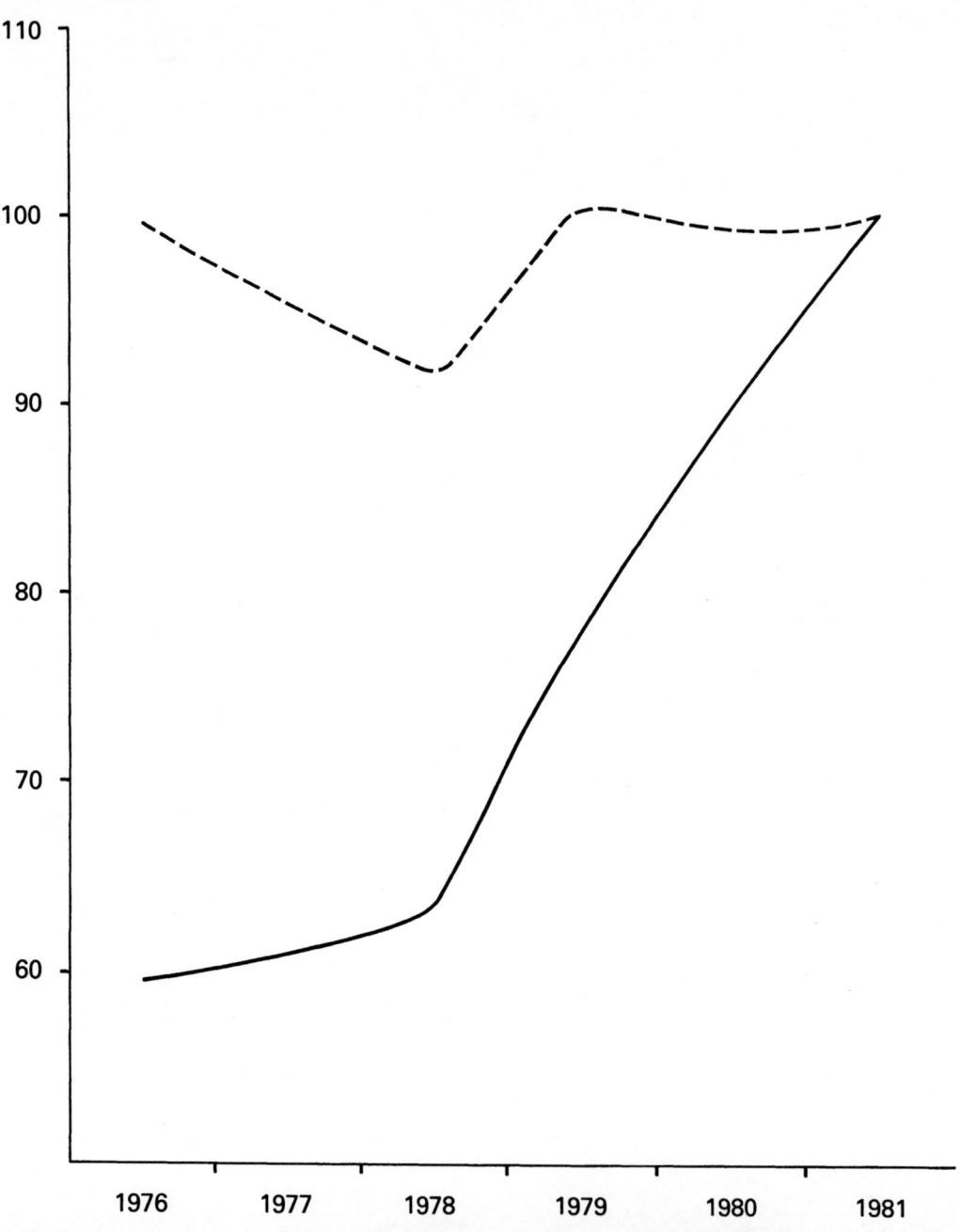

SUPPLY AND DEMAND BY MAIN MARKET AREA

	UK	EC (Ten)	Japan	USA
Production (1979/80 Averages) '000 tonnes K_2O	272	5067	43 (1979 only)	2232
Net Imports (1979/80 Averages) '000 tonnes All forms (K_2O)	330 (approx.)	1125 (approx.)	766 (1979 only)	5068
Potassium chloride (gross weight)				
Under 40% K_2O	..	29		
40-60% K_2O	248	1113		
Over 60% K_2O	325	813		
Potassium sulphate (gross weight)	19	5		
Source of Net Imports (%)	(KCl only)	(KCl only)		
Canada	45	3	50	94
European Community			15	
Spain	9	8		
United States		2		
East Germany	38	37		1
USSR	3	31	14	
Israel	4	18	5	4
Others	1		16	1
Net Exports (1979/80 Averages) '000 tonnes				
All forms (K_2O)	100 approx.	1000 approx.	6 (1979 only)	699

	UK	EC (Ten)	Japan	USA
Potassium chloride (gross weight)				
Under 40% K_2O	0.3	12		
40-60% K_2O	180	159		
Over 60% K_2O	1.4	870		
Potassium sulphate	..	846		
Consumption (1979/80 Averages) '000 tonnes K_2O	412 (fertilisers)	5000 (approx.)	736 (1979 only)	6672
Import Dependence				
Imports as % of consumption	80	23	100	76
Imports as % of consumption and net exports	64	19	100	69
Share of World Consumption %				
Total world	2	19	3	25
Consumption Growth % p.a.				
1970s	-0.7	1.5 (approx.)	1.0	4.5

****

RHENIUM

WORLD RESERVES
(tonnes of metal and % of total)

Developed			Less Developed			Centrally Planned			Total
Canada	320	(10.1)	Chile	1180	(37.2)	USSR	225	(7.1)	
USA	1180	(37.2)	Peru	180	(5.7)				
			Others	90	(2.8)				
Totals	1500	(47.2)		1450	(45.7)		225	(7.1)	3175

Rhenium is obtained as a by-product of molybdenite in porphyry copper operations. Identified world resources amount to 10,000 tonnes.

WORLD MINE PRODUCTION
(Kilograms and % of total 1979/80 Averages)

Note:- The available statistics are meagre and these figures are only approximate

Developed			Less Developed			Centrally Planned			Total
Canada	1900	(23)	Chile	4000	(48)	USSR	1150	(14)	
USA greater than	700	(8)	Peru	180	(2)	Others	270	(3)	
approx.			Others	200	(2)				
Totals	2500	(31)		4380	(52)		1420	(17)	8300

W Germany produced some 2000 Kilograms mainly from imported molybdenite concentrates (largely from Chile and to a lesser extent Canada).

RESERVE PRODUCTION RATIOS

Static reserve life (years)   : 400
Ratio of identified resources
to cumulative demand 1981-2000   66 : 1

CONSUMPTION

Statistics are only available for the United States, and even these are subject to a large margin of error.

United States consumption averaged 3810 Kilograms in 1979/80. It increased at an average compound rate of 7.2% per annum in the 1970s.

END USE PATTERNS 1980 (USA) %

Petroleum refining	92
Other	8

VALUE OF CONTAINED METAL IN ANNUAL PRODUCTION

$11 million (contained metal at average 1981 prices)

SUBSTITUTES

Non-rhenium catalysts are becoming more common. Iridium, gallium, germanium and silicon are among the metals being evaluated.

Substitutes in other applications are cobalt and tungsten for coatings on X-ray tubes, rhodium and rhodium-iridium for high temperature thermocouples, tungsten and platinum-ruthenium for coatings on electrical contacts and tungsten and tantalum for electron emitters.

TECHNICAL POSSIBILITIES

Use in high temperature applications such as some nickel-base alloys.

Radiation screens, semi conductors, resistors, small electromagnets, heat shields, diverse catalytic reactions are all possible new uses.

Changes in petroleum refining techniques to meet lower emission standards could have a detrimental effect on rhenium consumption.

PRICES

	1976	1977	1978	1979	1980	1981
US Metal Powder 99.99% $/lb	525	475	350	950	1500	675
Perrhenic acid $/lb	480	450	330	900	1400	550

Rhenium is a by-product of molybdenite which itself is recovered with or from porphyry copper ores. Production therefore is entirely dependent on Cu-Mo industry. Prices have risen rapidly in last decade due to increased demand from bimetallic catalyst industry. Dealer market.

MARKETING ARRANGEMENTS

Main sources of ore are Chile, Canada, US and USSR but recovery is concentrated in US, Germany, Sweden, Chile and USSR.

Index Numbers 1981 = 100

The solid line gives prices in money terms and the dotted line gives prices in 'real' 1981 terms

RHENIUM
US Metal Powder, 99·99%

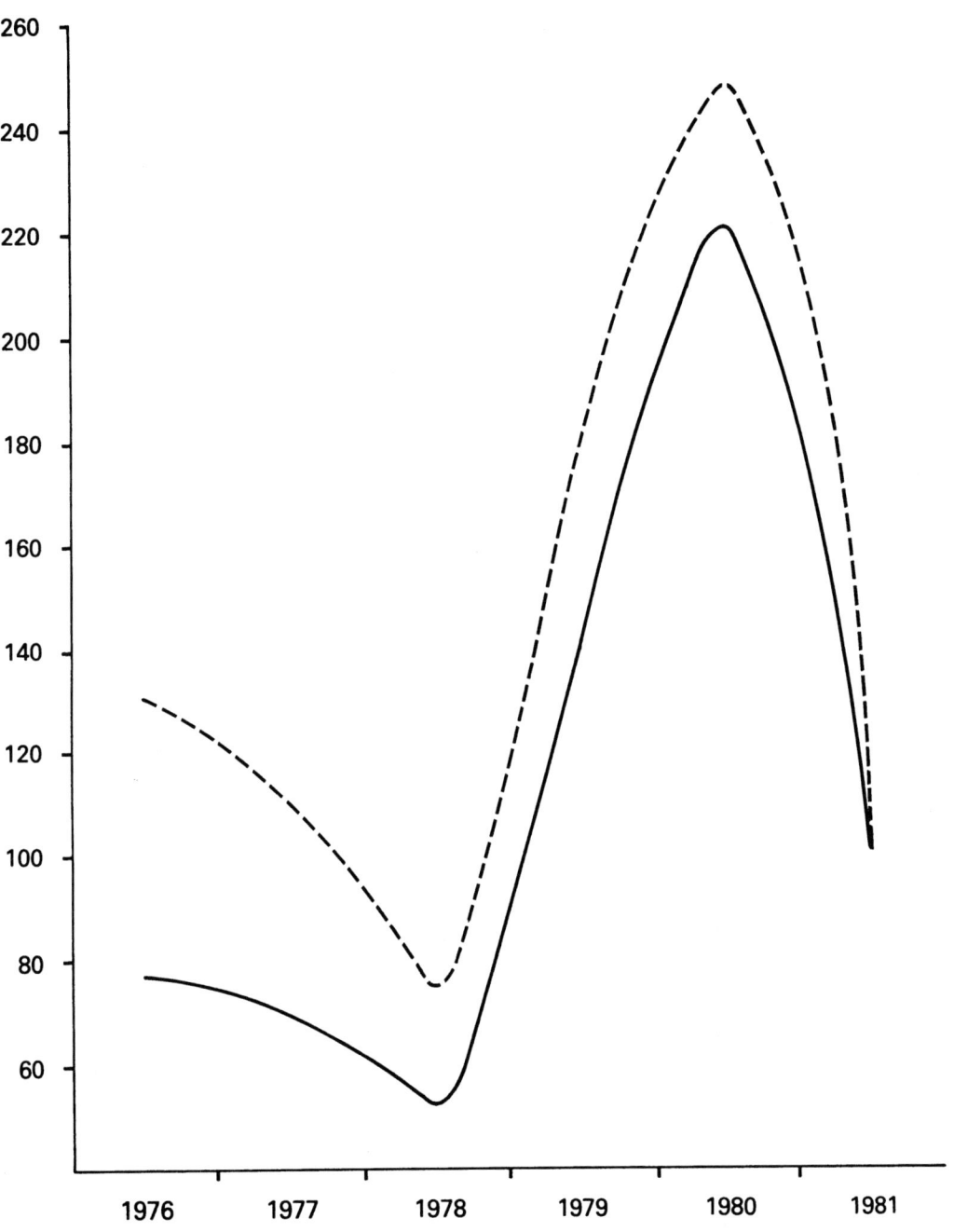

SUPPLY AND DEMAND BY MAIN MARKET AREA

	UK	EC (Ten)	Japan	USA
Production (1979/80 Averages) Kilograms				
Mine output	-	-	-	over 700
Refined output	negligible	2200 (approx.)	-	n/a
Net Imports (1979/80 Averages) Kilograms of contained rhenium				
Metal	c.300	n/a	n/a	327
Ammonium perrhenate	n/a	n/a	n/a	3014
Source of Net Imports (%) (Metal, perrhenic acid and ammonium perrhenate)				
European Community	Most	Figures not available (ores mainly from Canada, Chile & Peru)	Figures not available	53
USSR				2
Chile				43
Others				2
Net Exports (1979/80 Averages) Kilograms Metal	c.350	n/a	n/a	-
Consumption (1979/80 Averages) Kilograms contained rhenium	n/a	n/a	n/a	3810
Import Dependence				
Imports as % of consumption	100	100	100	88
Imports as % of consumption and net exports	100	100	100	88

	UK	EC (Ten)	Japan	USA
Share of World Consumption %				
Total world	n/a	n/a	n/a	46 (approx.)
Consumption Growth % p.a.				
1970s	n/a	n/a	n/a	7.2

****

SELENIUM

WORLD RESERVES
('000 tonnes of metal and % of total)

Developed			Less Developed			Centrally Planned			Total
Canada	20	(17)	Chile	21	(18)	Total	13	(11)	
USA	20	(17)	Mexico	6	(5)				
Others	5	(4)	Peru	7	(6)				
			Zaire	5	(4)				
			Zambia	7	(6)				
			Others	16	(12)				
Totals	45	(38)		62	(51)		13	(11)	120

Selenium occurs as a by-product with copper, and the above figures only cover the estimated content of economic copper deposits. Total identified world resources on this basis amount to some 400,000 tonnes. Substantially greater resources exist in association with other metals and coal deposits.

WORLD REFINERY PRODUCTION
(tonnes of metal and % of total 1979/80 Averages)

Selenium is recovered mainly from the anode slimes obtained from electrolytic refining of copper. Because the selenium content of copper ores varies widely it is impossible to estimate mine production accurately. The following figures cover refinery output.

Developed		Less Developed		Centrally Planned		Total
Australia	n/a	Chile	8	USSR	n/a	
Belgium	59	Mexico	78			
Canada	444	Peru	19			
Finland	17.5	Zambia	25			
W Germany	n/a					
Japan	515					
Sweden	68					
USA	204					
Yugoslavia	50					
Totals	1357.5		130			1487.5

Note:- Because the totals are incomplete no percentages are shown. Total world mine production has probably been around 1900-2000 tonnes and total refined output some 600 to 1700 tonnes in recent years.

Selenium is recovered from used electronic and photocopier components and recycled. Total North Amercian secondary production is some 50-100 tonnes per year, depending on the prices.

RESERVE/PRODUCTION RATIOS

Static reserve life (years)   : very large
Ratio of identified resources
to cumulative demand 1981-2000   7½ : 1

CONSUMPTION

	1979/80 Averages tonnes	% p.a. Growth rates 1970s
European Community	500-600 approx.	n/a
Japan	210	1.0
United States	357	-5.0

END USE PATTERNS 1980 (USA) %

Electronic and photocopier	
components	35
Glass manufacturing	30
Chemicals and Pigments	25
Other	10

VALUE OF CONTAINED METAL IN ANNUAL PRODUCTION

$20 million (refined metal at 1981 average prices)

SUBSTITUTES

Substitutes exist in most end uses. Organic chemicals are used in photocopying machines. Silicon substitutes in rectifier applications and cerium in glass manufacturing. Sulphur, lead, bismuth and tellurium may be used in steel applications.

TECHNICAL POSSIBILITIES

Quantity of selenium produced could decrease if hydro-metallurgical processes for leaching copper sulphide concentrates to recover copper are applied.

Increased recovery from flue dust and scrap. New users ustilising electrophotographic properties. Toxicity will limit use in pigments.

PRICES

	1976	1977	1978	1979	1980	1981
$/lb						
US major producer Commercial grade powder. min 99.5%	18	17.1	15	13.7	10.9	10.5*
European free market cif	14.6	13.3	11.6	11.2	8.5	4.5

* 5 months only. List price suspended due to market state.

Selenium is derived from anode slimes obtained from refining of copper and production is therefore independent of demand. There is both producer pricing and a dealer market with the former having almost ceased to exist in the current period of oversupply.

MARKETING ARRANGEMENTS

Canada and US are the two largest mine producers, Canada and Japan the two largest refinery producers. Selenium-Tellurium Development Association promotes interest in new uses of these two metals.

Index Numbers 1981 = 100

The solid line gives prices in money terms and the dotted line gives prices in 'real' 1981 terms

SELENIUM
European Free Market

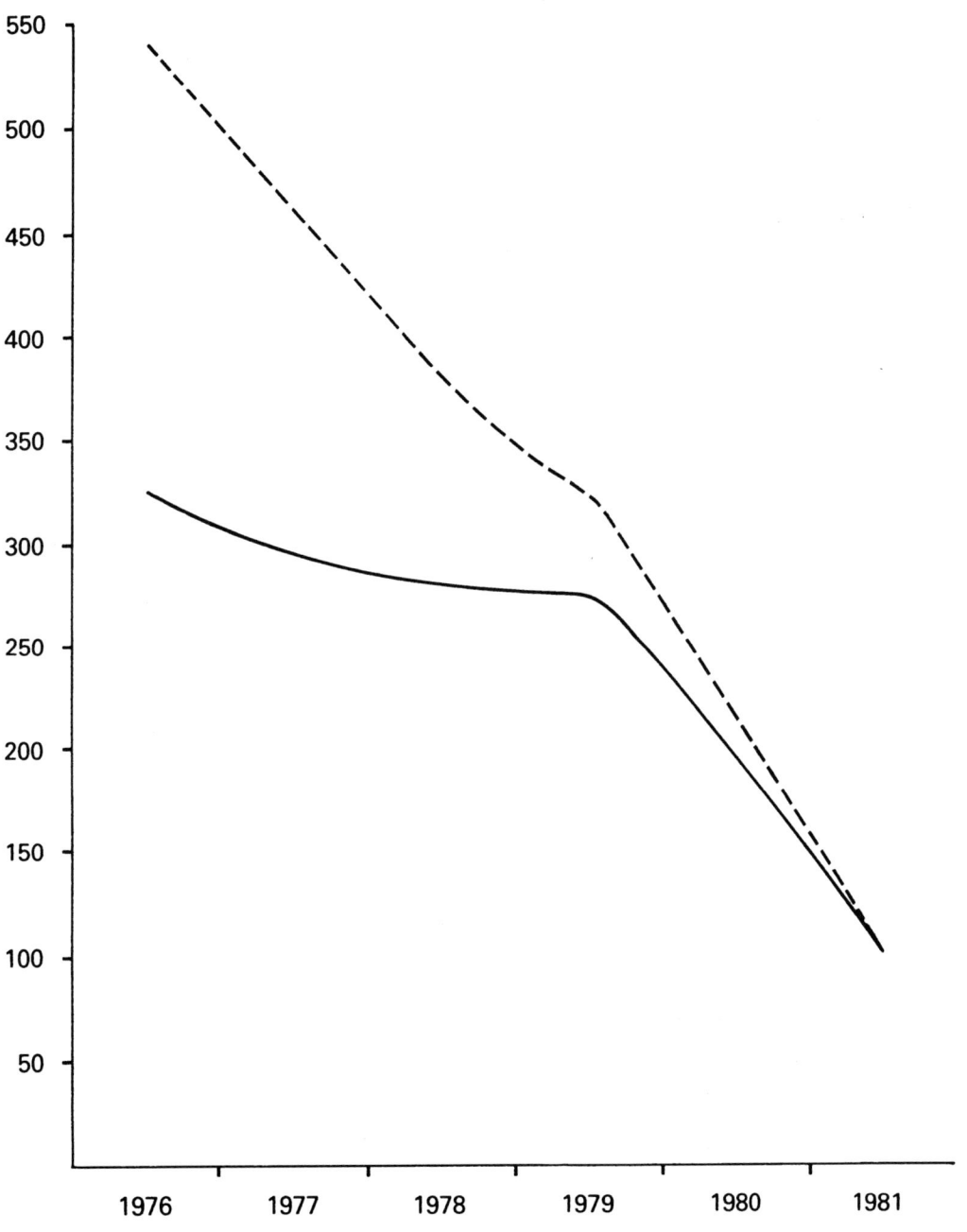

SUPPLY AND DEMAND BY MAIN MARKET AREA

	UK	EC (Ten)	Japan	USA
Production (1979/80 Averages) tonnes contained selenium				
Mine production	-	-	-	n/a
Refinery production	-	Belgium 59 W Germany n/a	515	203.6 (exc. secondary & mainly from domestic ores)
Net Imports (1979/80 Averages) tonnes contained selenium				
Metal and compounds	332.4	518.5	-	297
Source of Net Imports (%)				
Metal				
Canada	30	22		48
European Community	12			19
Finland		1		
Japan	16	30		15
Norway		2		
Sweden	6	6		3
United States	14	14		
Yugoslavia		4		8
Bulgaria		8		
Chile				5
Mexico	5	4		
Zambia		4		
Others	17	5		2
Net Exports (1979/80 Averages) tonnes contained selenium				
Metal	111.5	113	255.5	116.5
Consumption (1979/80 Averages) tonnes contained selenium	c.200	c.500-600	210.5	357

	UK	EC (Ten)	Japan	USA
Import Dependence				
Imports as % of consumption	100	100 (a)	100 (a)	83
Imports as % of consumption and net exports	100	100 (a)	100 (a)	63

(a) Allowing for import of original raw material, but partially or totally
self sufficient in refined output

Share of World Consumption %

| Western world (approx.) | 15 | c.33 to 40 | 14 | 25 |

Consumption Growth % p.a.

| 1970s | 2.6
(based on net
imports) | n/a | 1.0 | -5 |

****

SILICON

Silicon is an important constituent of quartzite and other sandstones.
There are ample reserves in most major producing countries in relation to
demand. Estimates of total reserves, and of their geographical
distribution, are not available.

WORLD PRODUCTION OF SILICON METAL
('000 tonnes and % of total 1979/80 Averages)

Developed			Less Developed			Centrally Planned			Total
Canada	27	(5.3)	Brazil	5	(1.0)	China	16	(3.1)	
France	52	(10.2)	India	1	(0.2)	Czecho-			
						slovakia	5	(1.0)	
W Germany	3	(0.6)				E Germany	4	(0.8)	
Italy	16	(3.1)				Hungary	2	(0.4)	
Japan	15	(3.0)				Poland	11	(2.2)	
Norway	63	(12.4)				USSR	57	(11.2)	
S Africa	33	(6.5)							
Spain	17	(3.3)							
Sweden	15	(3.0)							
Switzerland	9	(1.8)							
USA	126	(24.8)							
Yugoslavia	31	(6.1)							
Totals	407	(80.1)		6	(1.2)		95	(18.7)	508

WORLD PRODUCTION OF FERROSILICON
('000 tonnes and % of total 1979/80 Averages)

Developed			Less Developed			Centrally Planned			Total
Australia	20	(0.6)	Argentina	15	(0.4)	Bulgaria	16	(0.5)	
Canada	92	(2.7)	Brazil	88	(2.6)	China	185	(5.4)	
France	250	(7.3)	Chile	3	(0.1)	Czecho-slovakia	32	(0.9)	
W Germany	112	(3.3)	Colombia	1	(..)	E Germany	30	(0.9)	
Iceland	22	(0.6)	Egypt	5	(0.1)	Hungary	7	(0.2)	
Italy	76	(2.2)	India	60	(1.7)	N Korea	30	(0.9)	
Japan	323	(9.4)	Mexico	25	(0.7)	Poland	51	(1.5)	
Norway	328	(9.5)	Peru	1	(..)	Romania	25	(0.7)	
Portugal	23	(0.7)	Philippines	18	(0.5)	USSR	630	(18.3)	
S Africa	110	(3.2)	S Korea	30	(0.9)				
Spain	115	(3.3)	Taiwan	38	(1.1)				
Switzerland	5	(0.1)	Thailand	3	(0.1)				
Turkey	3	(0.1)	Venezuela	40	(1.2)				
USA	559	(16.3)							
Yugoslavia	67	(1.9)							
Totals	2105	(61.2)		327	(9.5)		1006	(29.3)	3438

The silicon content of ferrosilicon varies widely. The average in the USA in 1980 was 49%.

RESERVE/PRODUCTION RATIOS

For practical purposes so large as to be infinite.

CONSUMPTION

	1979/80 Averages '000 tonnes	% p.a. Growth rates 1970s
Silicon Metal		
European Community	122	n/a
Japan	73	10.3
United States	114	6.9
Ferrosilicon (gross weight)		
European Community	711	2 approx.
Japan	425	5
United States	451	0.8

END USE PATTERNS 1980 (USA) %

Transport	33
Machinery	20
Construction	14
Chemicals	12
Other	21

This covers the usage of silicon in all forms.

VALUE OF ANNUAL PRODUCTION

$4½ billion (at average 1981 prices of silicon and ferrosilicon)

SUBSTITUTES

Aluminium is among the alternatives for ferrosilicon as a deoxidiser in steel but at higher cost and production of side effects. Aluminium-silicon alloys can be replaced by some other aluminium alloys.

Germanium can be used in some semiconductor applications.

TECHNICAL POSSIBILITIES

Expansion of use in alloys particularly as a substitute for expensive additives such as chromium. Research in electronics is increasing use and demand for high purity silicon. Development of economically competitive silicon photovoltaic cells would increase demand here also.

Further development of high performance silicon-base ceramics are being developed as a substitute for superalloys and other metals in high temperature or high corrosive situations.

PRICES

	1976	1977	1978	1979	1980	1981
Metal UK 98% min £/tonne	510.8	567.5	501.3	625.1	651.9	625.0
Metal US 0.35% Fe/0.07% Ca ¢/lb	46.4	47.9	49.0	57.1	64.0	69.8
Ferrosilicon US Producer 75-77% Si ¢/lb	36.8	37.0	38.2	44.2	46.4	50.4

Prices mainly determined on contract basis of 3-6 months. Energy costs important.

MARKETING ARRANGEMENTS

Wide range of companies involved from integrated producers to one phase operators. Increasingly, location of ferrosilicon and silicon metal smelters in low power cost countries. Tendency also towards plant specialisation.

Index Numbers 1981 = 100

The solid line gives prices in money terms and the dotted line gives prices in 'real' 1981 terms

SILICON
US, Metal, 0·35% Fe/ 0·07% Ca

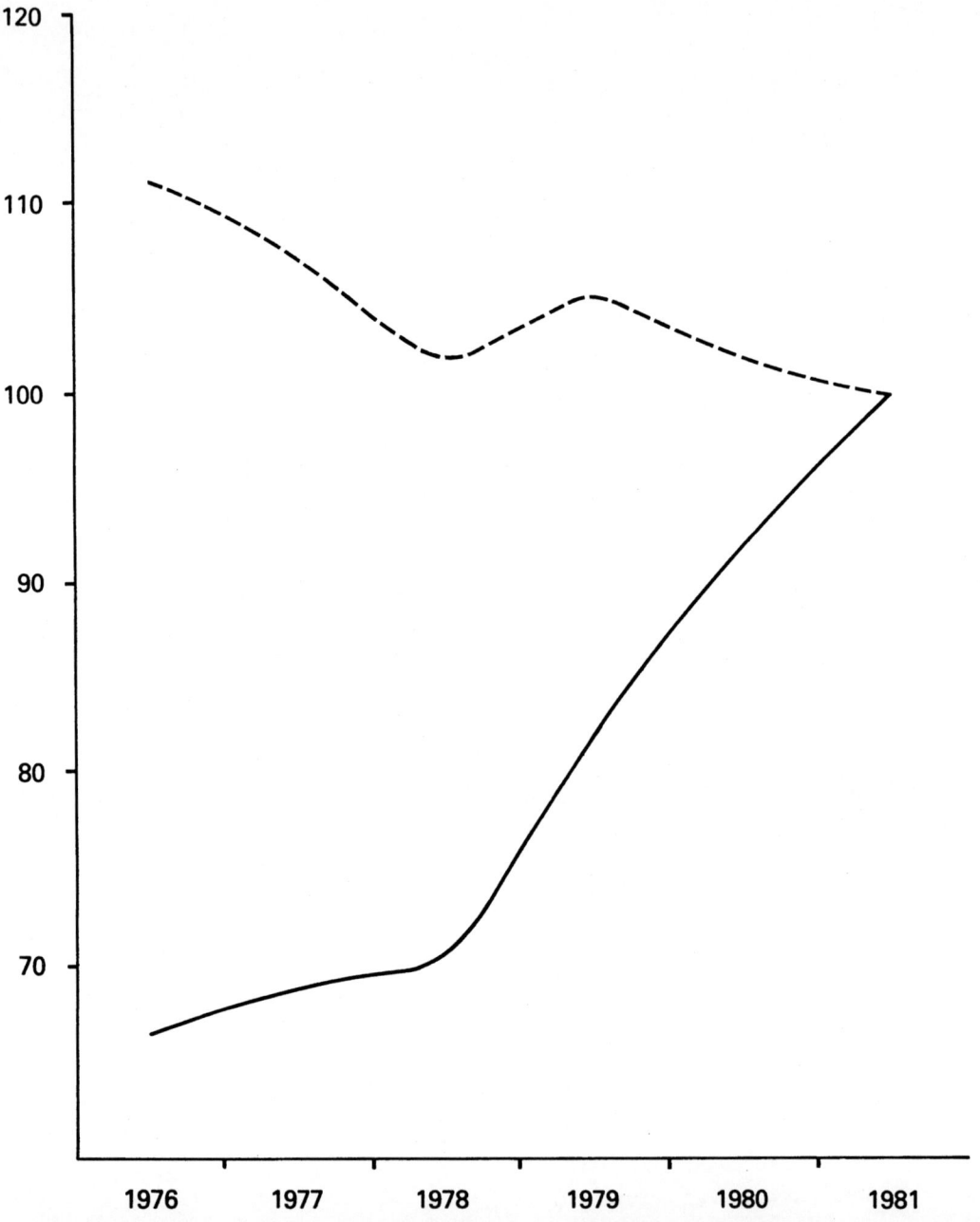

SUPPLY AND DEMAND BY MAIN MARKET AREA

	UK	EC (Ten)	Japan	USA
Production (1979/80 Averages) '000 tonnes				
Silicon metal	-	71.3	15.5	125.8
Ferrosilicon	-	437.7	323.4	559.3
Net Imports (1979/80 Averages) '000 tonnes				
Silicon metal	24.0	62.2	58.9	22.4
Ferrosilicon	94.1	307.2	127.8	83.8
Source of Net Imports (%)				
Silicon Metal				
Canada		6	13	49
European Community	39		11	3
Norway	40	39	8	9
Portugal	5	10	13	
S Africa	6	15	15	23
Spain		4	16	
Sweden	6	7	3	
Switzerland		10		
United States			12	
Yugoslavia		8	5	14
Brazil			4	
Others	4	1	-	2
Ferrosilicon				
Canada			10	23
European Community	6		7	11
Norway	67	64	27	24
Portugal		3		
S Africa	3	2	13	6
Spain	10	15	5	
Yugoslavia		3	5	
Bulgaria		2		
Romania	2	2		
USSR	2	2		

	UK	EC (Ten)	Japan	USA
Brazil			1	18
India			7	
Philippines			9	
Venezuela			8	11
Others	10	7	8	7

Net Exports (1979/80 Averages) '000 tonnes

	UK	EC (Ten)	Japan	USA
Silicon metal	0.4	4.5	-	8.8
Ferrosilicon	1.5	26.4	2.3	22.6

Consumption (1979/80 Averages) '000 tonnes

	UK	EC (Ten)	Japan	USA
Silicon metal	23.6	121.9	73.5	113.6
Ferrosilicon	92.6	711.3	425	450.6

Import Dependence

Imports as % of consumption

	UK	EC (Ten)	Japan	USA
Silicon metal	100	51	80	20
Ferrosilicon	100	43	30	19

Imports as % of consumption and net exports

	UK	EC (Ten)	Japan	USA
Silicon metal	100	49	80	18
Ferrosilicon	100	42	30	18
All forms (si content)	100	44	43	20

Share of World Consumption %

Total world (approx.)

	UK	EC (Ten)	Japan	USA
Silicon metal	5	24	14	22
Ferrosilicon	3	21	12	13

Consumption Growth % p.a.

	UK	EC (Ten)	Japan	USA
1970s total	-2	n/a	5.8	2.4
of which:- Silicon metal	2.9	n/a	10.3	6.9
Ferrosilicon	-5.2	2 (approx.)	5	0.8

****

SILVER

WORLD RESERVES
('000 tonnes and % of total)

Developed			Less Developed			Centrally Planned			Total
Australia	32	(12.5)	Mexico	33	(12.9)	USSR	50	(19.6)	
Canada	50	(19.6)	Peru	19	(7.5)	Others	12	(4.7)	
USA	47	(18.4)	Others	10	(3.9)				
Others	2	(0.8)							
Totals	131	(51.4)		62	(24.3)		62	(24.3)	255

Identified world resources are estimated at 770,000 tonnes. The greater
part of reserves and resources is associated with base metals such as
copper, lead and zinc.

WORLD MINE PRODUCTION
(tonnes of metal and % of total 1979/80 Averages)

Developed			Less Developed			Centrally Planned			Total
Australia	801	(7.5)	Bolivia	184	(1.7)	Poland	734	(6.9)	
Canada	1092	(10.2)	Chile	285	(2.7)	USSR	1550	(14.5)	
Japan	273	(2.6)	Mexico	1505	(14.1)	Others	234	(2.2)	
S Africa	99	(0.9)	Morocco	100	(0.9)				
Spain	137	(1.3)	Peru	1282	(12.0)				
Sweden	169	(1.6)	Others	731	(6.8)				
USA	1079	(10.1)							
Yugoslavia	156	(1.5)							
Others	256	(2.4)							
Totals	4062	(38.0)		4087	(38.3)		2528	(23.7)	10677

WESTERN WORLD SILVER SUPPLIES
tonnes of metal

	1978	1979	1980
Mine Production	8351	8398	7931
Secondary Source of Supply			
US Treasury	3	3	3
Other governments	261	96	162
Demonetized coin	435	793	1711
Indian stocks	1415	1042	1297
Salvage and other miscellaneous sources	3001	2504	3779
Liquidation of (additions to) private bullion stocks	1429	1086	(3813)
Total other supplies	6544	5524	3139
Available for world consumption	14895	13922	11070

Source: The Silver Market 1980 Handy & Harman

RESERVE/PRODUCTION RATIOS

Static reserve life (years)   24
Ratio of identified resources
to cumulative primary demand 1981-2000 2.7 : 1

This ignores substantial secondary recovery and above ground stocks.

CONSUMPTION

	1979/80 Averages tonnes	% p.a. Growth rates 1970s
Industrial uses		
European Community (a)	3809	-1.9
Japan	2028	3.9
United States	4306	0.8
Other Countries	1676	5.2
Total Industrial usage	11819	1.1
Coinage	677	-2.6
Total Consumption	12496	0.8

Source: Handy & Harman reports on silver

(a) Belgium, France, W Germany, Italy and United Kingdon only

END USE PATTERNS 1980 (USA) %

This covers only industrial uses and excludes 'investment' demand.

Photography	35
Electrical and electronic components	29
Sterlingware and electroplated ware *	10
Brazing alloys and solders	8
Other	18

* A higher percentage in less industrialised countries.

VALUE OF CONTAINED METAL IN ANNUAL PRODUCTION

$3.6 billion (mine production at average 1981 prices)

SUBSTITUTES

Stainless steel is an economic alternative in table flatware. Aluminium and rhodium are used for reflecting surfaces. Tantalum is a substitute for surgical plates, fins and sutures.

Silver has been replaced in coinage in many countries by cupro-nickel, cupro-zinc, nickel and aluminium.

TECHNICAL POSSIBILITIES

Improvements in solid-state switching and in electroplating and cladding technology will extend life of electronic equipment, decreasing demand.

New processing techniques may improve recovery.

Development of silver-free film. Introduction of new types of silver-containing batteries.

PRICES

	1976	1977	1978	1979	1980	1981
LME Cash $/troy oz	4.39	4.64	5.41	10.95	21.10	10.42
Handy and Harman New York $/troy oz	4.35	4.62	5.40	11.09	20.63	10.52
LME Cash Monthly Range £/troy oz	2.00-2.71	2.55-2.83	2.56-3.00	3.10-10.12	5.46-17.47	4.44-6.12

Result of interaction of supply and demand with variable, and sometimes considerable, speculative activity (e.g. 1980).

MARKETING ARRANGEMENTS

Approximately 50% of new mined silver is from predominantly silver ores, the balance being derived as by-product of copper, lead and zinc. Demand currently exceeds primary supply and the deficit is supplied by secondary sources; these rose to such a high level in 1980/81 that total supply considerably exceeded demand with consequent depressing effect on prices. US stockpile excesses overhang the market.

Index Numbers 1981 = 100

The solid line gives prices in money terms and the dotted line gives prices
in 'real' 1981 terms

SILVER
LME Cash

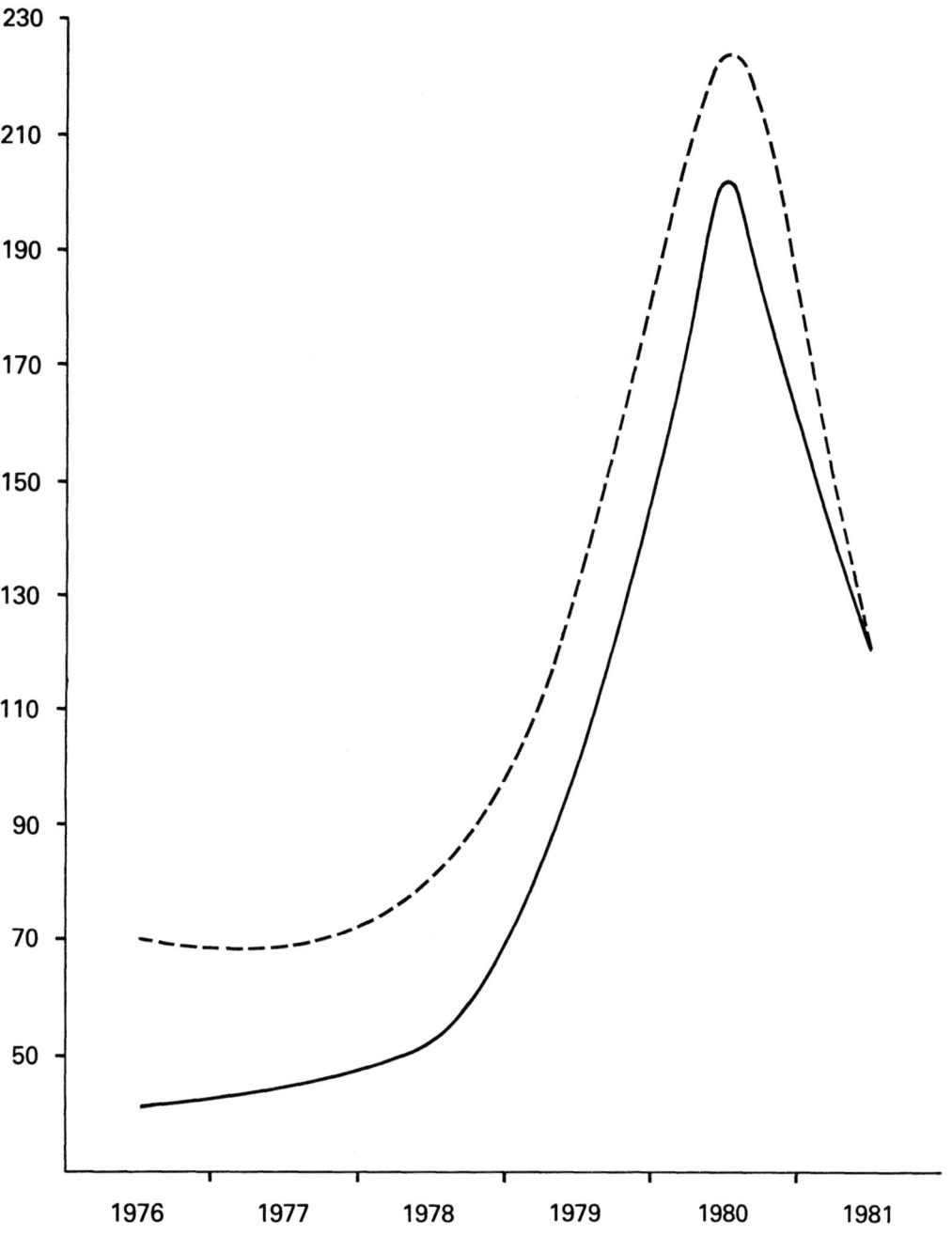

SUPPLY AND DEMAND BY MAIN MARKET AREA

	UK	EC (Ten)	Japan	USA
Production (1979/80 Averages) tonnes				
Mine production	0.6	243.7	273.5	1078.9
Refinery production : New	n/a	n/a	n/a	1401
Secondary (old scrap)	n/a	n/a	n/a	1444
Net Imports (1979/80 Averages) tonnes				
Ores and concentrates	-	-	585	305.2
Semi processed (inc. scrap)	646.4)	4139.5	-	129.4
Refined (inc. partly worked)	2152.9)		626.4	2227.5
of which A		2550		
B		1364.5		
C		225		

A   Crude and semi processed 99.9% or more
B   Other crude and semi processed
C   Wrought

Source of Net Imports (%)

Ores and concentrates

	UK	EC (Ten)	Japan	USA
Bolivia	-	-	24	-
S Korea	-	-	76	-

Metal				(All forms)
Australia	3	2	7	
Canada			1	34
European Community	35		1	15
Japan		2		
S Africa	3	2		
Spain		3		
Sweden		4		
Switzerland	2	15		
United States	21	33	5	
Yugoslavia				1
E Germany	5	3		
N Korea		2		
Poland		6		

	UK	EC (Ten)	Japan	USA
Chile		2		2
India	2			
Mexico	7	7	48	23
Peru		1	31	22
S Korea			5	
United Arab Emirates	8	5		
Others	14	13	2	3

Net Exports (1979/80 Averages)
tonnes

	UK	EC (Ten)	Japan	USA
Ores and concentrates	-	-		5.5
Semi processed (inc. scrap)	424.6)	3155		661.3
Refined (inc. partly worked)	2158.8)		89	1143.6

Consumption (1979/80 Averages)
tonnes

	UK	EC (Ten)	Japan	USA
Industrial	731	4000	2024.8	4385
Coinage		(approx.)		3.7

Import Dependence

	UK	EC (Ten)	Japan	USA
Imports as % of consumption (exc. coinage)	100	100	60	61
Imports as % of consumption and net exports (exc. coinage)	100	58 (approx.)	57	43

Share of World Industrial Consumption %

	UK	EC (Ten)	Japan	USA
Western world	6	over 32	17	36

Consumption Growth % p.a.

	UK	EC (Ten)	Japan	USA
1970s	-0.7	-1.9	3.9	0.8

Note:- Some of the figures in this table (e.g. on consumption) differ slightly from those of earlier tables, derived from different sources.

****

SULPHUR

WORLD RESERVES
('000 tonnes and % of total)

Developed			Less Developed			Centrally Planned			Total
Canada	250	(14.2)	Iraq	150	(8.5)	China	25	(1.4)	
France	30	(1.7)	Mexico	90	(5.1)	Poland	150	(8.5)	
W Germany	30	(1.7)	Near East	250	(14.2)	USSR	250	(14.2)	
Italy	15	(0.8)	Others	105	(5.9)	Others	60	(3.4)	
Japan	10	(0.6)							
Spain	30	(1.7)							
USA	175	(9.9)							
Others	145	(8.2)							
Totals	685	(38.8)		595	(33.7)		485	(27.5)	1765

Identified world resources total 6,385 million tonnes.

WORLD PRODUCTION IN ALL FORMS
(million tonnes and % of total 1979/80 Averages)

A significant percentage of output is a by-product of metallurgical operations or petroleum refining.

Developed			Less Developed			Centrally Planned			Total
Canada	7216	(13.0)	Iraq	640	(1.2)	China	2300	(4.1)	
Finland	442	(0.8)	Mexico	2463	(4.4)	Poland	5020	(9.0)	
France	2133	(3.8)	Saudi Arabia	413	(0.7)	Romania	570	(1.0)	
W Germany	1806	(3.2)	Others	1511	(2.7)	USSR	10725	(19.3)	
Italy	587	(1.1)				Others	1110	(2.0)	
Japan	2897	(5.2)							
S Africa	455	(0.8)							
Spain	1196	(2.1)							
USA	11970	(21.5)							
Others	2186	(3.9)							
Totals	30888	(55.5)		5027	(9.0)		19725	(35.5)	55640

Of the total output just over 25% was Frasch, 6½% was native sulphur and 18% was from pyrites. The balance came from by-product sources.

RESERVE/PRODUCTION RATIOS

Static reserve life (years)   32
Ratio of identified resources
to cumulative demand 1981-2000   3.4 : 1

CONSUMPTION

Sulphur in all forms

	1979/80 Averages '000 tonnes	% p.a. Growth rates 1970s
European Community (a)	8208	0.7
Japan	2454	-1.4
United States	13862	3.4
Other Countries	13846	5.5
Total Western world	38370	3.0
Centrally Planned Economies	16313	n/a
Total world	54683	n/a

Source: British Sulphur Corporation Statistics

(a) Belgium, France, W Germany, Italy, Netherlands, United Kingdom and Greece only

END USE PATTERNS 1980 (USA) %

Fertilisers	60
Other chemical products	8
Metal mining	4
Petroleum refining	7
Other uses	21

VALUE OF ANNUAL PRODUCTION

$7.7 billion (at average 1981 prices)

SUBSTITUTES

Current and predicted price levels deter substitution except for acids in certain applications e.g. paints.

TECHNICAL POSSIBILITIES

Substitution of sulphur for up to 50% of asphalt in asphaltic road coverings.

Development of specialised sulphur concrete materials.

PRICES

	1976	1977	1978	1979	1980	1981
US Frasch, export bright, ex-terminal Holland $/lt (long term contracts)	65	66	66	79	113	136
Liquid sulphur contracts North West Europe delivered, ex-terminal range $/tonne			57-60*	67-85	103-135	139-152.5

* Second half only

Producer pricing for long term contracts. Transport costs very important.

MARKETING ARRANGEMENTS

Approximately 50% of world production is from countries in which the industry is nationalised (e.g. USSR, Mexico) or in which the governments have partial ownership (e.g. France, Spain) or exercise some measure of control (e.g. Japan). Production is worldwide with elemental sulphur (frasch and brimstone) accounting for 72% of W world production, pyrites 11% and smelter gases 16%. Supply pattern likely to be restructured by 2000, as production from primary sources is phased out in favour of co-product sources (coal, petroleum, natural gas, metal smelting) which serve to control environmental problems but at increased capital and operating costs.

Index Numbers 1981 = 100

The solid line gives prices in money terms and the dotted line gives prices in 'real' 1981 terms

SULPHUR
US Frasch, export

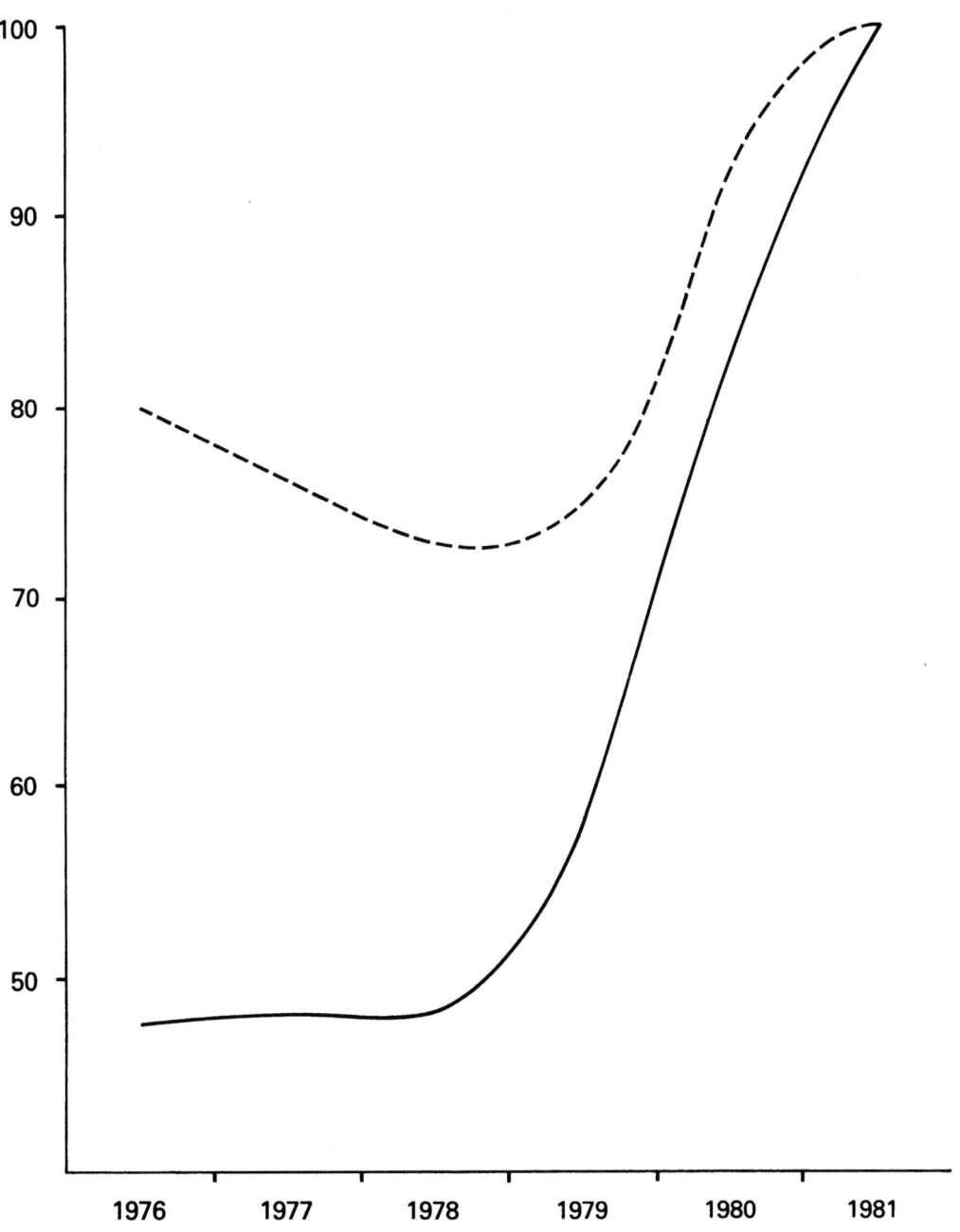

SUPPLY AND DEMAND BY MAIN MARKET AREA

	UK	EC (Ten)	Japan	USA
Production (1979/80 Averages) '000 tonnes				
Sulphur in all forms	125.5	5209.5	2819	12646.5
Net Imports (1979/80 Averages) '000 tonnes	1167.4	2810	nil (contained in non-ferrous ores etc.)	2508.5
Source of Net Imports (%)				
Canada	11	26	-	55
European Community	46			
United States	1	25		
Poland	33	43		
Mexico	4	3		44
Others	5	3		1
Net Exports (1979/80 Averages) '000 tonnes	1.2	737.6	389.5	1818
Consumption (1979/80 Averages) '000 tonnes				
Sulphur in all forms	1268	8208 (exc. Denmark and Ireland)	2454	13862
Import Dependence				
Imports as % of consumption	92	34	-	18
Imports as % of consumption and net exports	92	31	-	16
Share of World Consumption %				
Western world	3	21	6	36
Total world	2	15	4	25
Consumption Growth % p.a.				
All forms 1970s	-0.7	0.7	-1.4	3.4

****

TANTALUM

WORLD RESERVES
(tonnes of metal and % of total)

Developed			Less Developed			Centrally Planned			Total
Australia	13600	(17.7)	Brazil	3200	(4.2)	USSR	4500	(5.8)	
Canada	910	(1.2)	Malaysia	3630	(4.7)				
Others	500	(0.6)	Nigeria	7250	(9.4)				
			Thailand	4540	(5.9)				
			Zaire	37200	(48.3)				
			Others	1670	(2.2)				
Totals	15010	(19.5)		57490	(74.7)		4500	(5.8)	77000

Identified world resources exceed 265,000 tonnes.

WORLD MINE PRODUCTION
(tonnes of contained metal 1979/80 Averages)

Developed		Less Developed		Centrally Planned		Total
Australia	60	Brazil	122	Total	24	
Canada	117	Malaysia	3			
Portugal	1	Mozambique	30			
Spain	n/a	Namibia	n/a			
		Nigeria	34			
		Rwanda	9			
		Thailand	77			
		Uganda less than	1			
		Zaire	9			
		Zambia	n/a			
		Zimbabwe	8			
Totals	178		292		24	494

This table excludes production of tantalum from tin slags which is concentrated in Malaysia (greater than 100), Nigeria (c.60), Thailand (greater than 125), Zaire (greater than 20), USSR (greater than 100), with other countries producing small quantities. Excluding the USSR, world tantalum production from tin slags was about 450 tonnes in 1979. This means that total world production of tantalum from all sources averaged roughly 1050 tonnes in 1979/80.

Because tin slags are excluded no percentage shares of world output are given in the table.

RESERVE/PRODUCTION RATIOS

Static reserve life (years)   approx. 75
Ratio of identified resources
to cumulative demand 1981-2000   9 : 1

CONSUMPTION

	1979/80 Averages tonnes	% p.a. Growth rates 1970s
European Community	c.300	n/a
Japan	140 to 150	13.6 (powder only)
United States	596	1.8

END USE PATTERNS 1980 (USA) %

Electronic components	73
Machinery	19
Transport	6
Other	2

VALUE OF CONTAINED METAL IN ANNUAL PRODUCTION

$100 million (at average 1981 prices)

SUBSTITUTES

Aluminium and ceramics compete in capacitors; silicon, germanium and selenium are alternatives in rectifiers; zirconium, titanium can substitute as getters in electronic tubes and in corrosion-resistant equipment.

Columbium can replace tantalum in some carbides, and, along with platinum, in corrosion-resistant equipment and high temperature uses. Glass and titanium are further alternatives in corrosion-resistant equipment, molybdenum and tungsten compete in high temperature applications.

TECHNICAL POSSIBILITIES

Improved concentration, beneficiation and extraction methods.

Reduction in quantity of tantalum required per capacitor through higher capacitance ratings.

Increased use in superalloys.

PRICES

	1976	1977	1978	1979	1980	1981
Tanco Tantalite 50% Ta_2O_5						
$/lb Ta_2O_5	16	19.3	26.5	60.1	97.7	93.8
Spot Tantalite Ore cif US ports 60% combined Cb_2O_5						
Range of prices $/lb Ta_2O_5	15.5-17.25	17.25-26.5	22.75-41	38-105	90-120	35-108

Large producers have list prices, but dealer market important.

MARKETING ARRANGEMENTS

Production divided between mining of ores e.g. Canada, Australia, and processing of tin slags e.g Thailand. Some countries combine both methods e.g. Brazil. Substantial by-product production, usually with columbium. Largest producer is Tantalum Mining Corp. of Canada (Tanco). USA currently adding to their tantalum stockpile.

Index Numbers 1981 = 100

The solid line gives prices in money terms and the dotted line gives prices in 'real' 1981 terms

TANTALUM
Tanco Tantalite, 50% Ta_2O_5

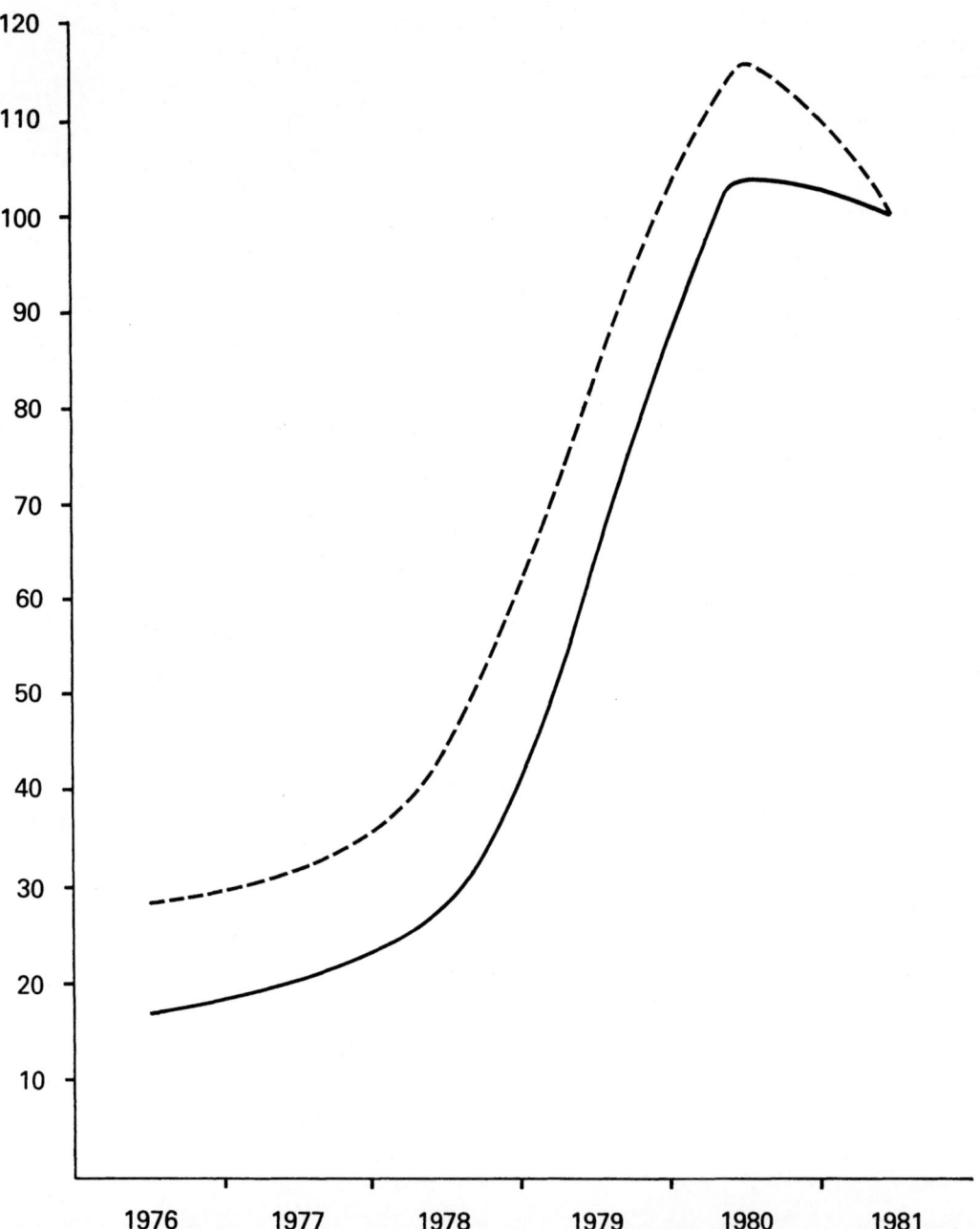

SUPPLY AND DEMAND BY MAIN MARKET AREA

	UK	EC (Ten)	Japan	USA
Production (1979/80 Averages) tonnes ta content				
Ores and concentrates	-	-	-	-
Net Imports (1979/80 Averages) tonnes ta content				
Mineral concentrates	inc. with niobium	inc. with niobium	107 (a)	338
Metal and alloys inc. powder	67	277.5	70.3	64
Tin slags	n/a	n/a	n/a	517 (1979 only)

(a) Gross weight 306.5 (the 35% assumed ta content is that of US imports in 1980)

Source of Net Imports (%)

Ores and concentrates

	UK	EC (Ten)	Japan	USA
Australia	n/a	n/a	5	15
Canada	n/a	n/a	6	29
European Community	n/a	n/a	12	11
Spain	n/a	n/a		3
Brazil	n/a	n/a	5	19
Malaysia	n/a	n/a	44	4
Rwanda	n/a	n/a		6
Thailand	n/a	n/a	27	3
Zaire	n/a	n/a		5
Others	n/a	n/a	1	5

Metal etc.

	UK	EC (Ten)	Japan	USA
Austria	3	4		
Canada				11
European Community	41		10	54
Japan		4		
United States	55	86	85	

	UK	EC (Ten)	Japan	USA
Mexico				27 (waste and scrap)
Others	1	6	5	8
Tin Slags	n/a	n/a	n/a	mainly Malaysia & Thailand
Net Exports (1979/80 Averages) tonnes ta content				
Concentrate, metal, alloys, waste and scrap	38.5	27.5	31.5	324
Consumption (1979/80 Averages) tonnes ta content	c.30	c.250 to 300	Total 140 to 150 Powder 99	596
Import Dependence				
Imports as % of consumption	100	100	100	100
Imports as % of consumption and net exports	100	100	100	100
Share of World Consumption %				
Western world	c.2	c.25	c.15	c.55
Consumption Growth % p.a.				
1970s	n/a	n/a	13.6 (powder only)	1.8

****

TIN

WORLD RESERVES
('000 tonnes of metal and % of total)

Developed			Less Developed			Centrally Planned			Total
Australia	350	(3.6)	Bolivia	980	(9.9)	China	1500	(15.2)	
Canada	20	(0.2)	Brazil	400	(4.1)	USSR	1000	(10.1)	
Portugal and									
Spain	30	(0.3)	Burma	500	(5.1)				
S Africa	(150)	(1.5)	Indonesia	1550	(15.7)				
UK	260	(2.6)	Malaysia	1200	(12.2)				
Others	60	(0.6)	Nigeria	280	(2.8)				
			Thailand	1200	(12.2)				
			Zaire	200	(2.0)				
			Others	190	(1.9)				
Totals	870	(8.8)		6500	(65.9)		2500	(25.3)	9870

Total identified world resources are estimated at 37 million tonnes.

WORLD MINE PRODUCTION
(tonnes of contained tin and % of total 1979/80 Averages)

Developed			Less Developed			Centrally Planned			Total
Australia	11481	(4.9)	Argentina	500	(0.2)	China	16500	(7.0)	
Canada	288	(0.1)	Bolivia	27526	(11.7)	Czecho-			
						slovakia	180	(0.1)	
Japan	604	(0.3)	Brazil	6787	(2.9)	E Germany	1700	(0.7)	
Portugal	250	(0.1)	Burma	1050	(0.4)	USSR	17500	(7.4)	
S Africa	2563	(1.1)	Indonesia	30984	(13.1)	Vietnam	300	(0.1)	
Spain	472	(0.2)	Laos	600	(0.3)				
UK	2701	(1.1)	Malaysia	62200	(26.3)				
USA	96	(..)	Namibia	1000	(0.4)				
			Nigeria	2638	(1.1)				
			Peru	1076	(0.5)				
			Rwanda	1500	(0.6)				
			Uganda	120	(0.1)				
			Thailand	33823	(14.3)				
			Zaire	3229	(1.4)				
			Zimbabwe	940	(0.4)				
			Other and unspecified Asian	7492	(3.2)				
Totals	18455	(7.8)		181465	(76.9)		36180	(15.3)	236100

WORLD PRODUCTION OF PRIMARY METAL
('000 tonnes of metal and % of total 1979/80 Averages)

Developed			Less Developed			Centrally Planned			Total
Australia	5.1	(2.2)	Argentina	0.1	(..)	China	16.0	(6.8)	
Belgium	2.5	(1.1)	Bolivia	16.6	(7.1)	E Germany	1.7	(0.7)	
W Germany	1.6	(0.7)	Brazil	9.5	(4.0)	USSR	17.5	(7.4)	
Japan	1.3	(0.6)	Indonesia	29.1	(12.4)				
Netherlands	1.4	(0.6)	S Korea	0.4	(0.2)				
Portugal	0.4	(0.2)	Malaysia	72.2	(30.7)				
S Africa	1.9	(0.8)	Mexico	0.5	(0.2)				
Spain	3.8	(1.6)	Nigeria	2.8	(1.2)				
UK	6.9	(2.9)	Singapore	4.0	(1.7)				
USA	4.2	(1.8)	Thailand	33.9	(14.4)				
			Zaire	0.7	(0.3)				
			Zimbabwe	0.9	(0.4)				
Totals	29.1	(12.4)		170.7	(72.6)		35.2	(15.0)	235.1

WORLD PRODUCTION OF SECONDARY METAL
('000 tonnes of metal and % of total 1979/80 Averages)

Developed		Less Developed		Centrally Planned		Total
Australia	0.5	Brazil	0.2	Czechoslovakia	0.1	
Canada	0.2	India	0.1	Others	n/a	
W Germany	1.6	Thailand	0.1			
Portugal	0.6					
UK	4.5					
USA	1.7					
Other Europe	0.7					
Totals	9.8		0.4		0.1	10.3

This table is incomplete.

RESERVE/PRODUCTION RATIOS

Static reserve life (years)   42
Ratio of identified resources to
cumulative 1981-2000 consumption 7 : 1

CONSUMPTION OF METAL

	1979/80 Averages '000 tonnes	% p.a. Growth rates 1960-70	1970-80
European Community (Ten)	51.7	-0.1	-1.7
Japan	31.0	6.9	1.9
United States	49.1	1.3	-1.6
Others	42.3	1.4	2.3
Total Western world	174.1	1.4	-0.2
Total world	228.8	2.1	-0.4

END USE PATTERNS 1980 (USA) %

Cans and containers	29*
Electrical	15
Construction	15
Transport	12
Other	29

* Canning's share is more than 50% in the world as a whole.

VALUE OF CONTAINED METAL IN ANNUAL PRODUCTION

$3.3 billion (primary refined metal at average 1981 prices)

SUBSTITUTES

Aluminium, tin-free steel, glass, paper, plastics all compete with tin in cans.

Non-metallic materials, copper, aluminium and zinc-coating products are alternatives in roofing and construction applications.

Aluminium alloys, copper-base alloys and plastics can substitute in bronze.

Other chemicals can replace tin compounds for use as fungicides and biocides or polyvinyl chloride stabilisers.

Epoxy resins can be used for solder though not as effectively.

Babbit metal can be replaced by low tin aluminium, copper, or lead bearing alloys and roller or ball bearings.

TECHNICAL POSSIBILITIES

Increasing recovery from slimes in beneficiation stage.

PRICES

	1976	1977	1978	1979	1980	1981
LME Standard Grade Cash $/lb	3.45	4.92	5.85	6.81	7.62	6.43
Penang $/lb	3.39	4.86	5.68	6.72	7.46	6.38
Penang $M/Kg	18.89	26.22	28.78	32.32	35.62	32.34
LME Monthly Average Range £/tonne	3074-5001	5274-6906	5942-7678	6614-7719	6266-7920	5939-8376

Most tin trade is related to Penang or LME market determined prices, as modified by ITC Intervention. Wide cost range within and between countries depending on production method. Malaysian gravel pumps (15% of output) marginal producers. Costs, net of taxes, well below prices, but taxes almost a 'cost'. Bolivian mines the highest cost. Tin prices theoretically subject to ITC intervention levels but ceiling often breached. Prices have recently been influenced by a producer group buying all available tin.

The ITC price ranges have been altered periodically to keep them roughly in line with cost changes. The chart shows how they have moved.

MARKETING ARRANGEMENTS

Production mainly in developing countries. State control in Bolivia. Tendency recently to smelt ores at source. International Tin Council, consisting of producers and consumers, overseas market. Sixth International Tin Agreement came into provisional operation on 1st July 1982 but dogged by political differences over size of buffer stock and export controls. Bolivia, USA and USSR who were members of Fifth Agreement have not joined. Malaysia is pressing for the establishment of a separate Association of Tin Producing Countries. Main problems are buffer stock size and export controls. The US is currently selling 35000 t of tin from its stockpile to international markets.

Index Numbers 1981 = 100

The solid line gives prices in money terms and the dotted line gives prices in 'real' 1981 terms

TIN
LME Standard Grade, Cash

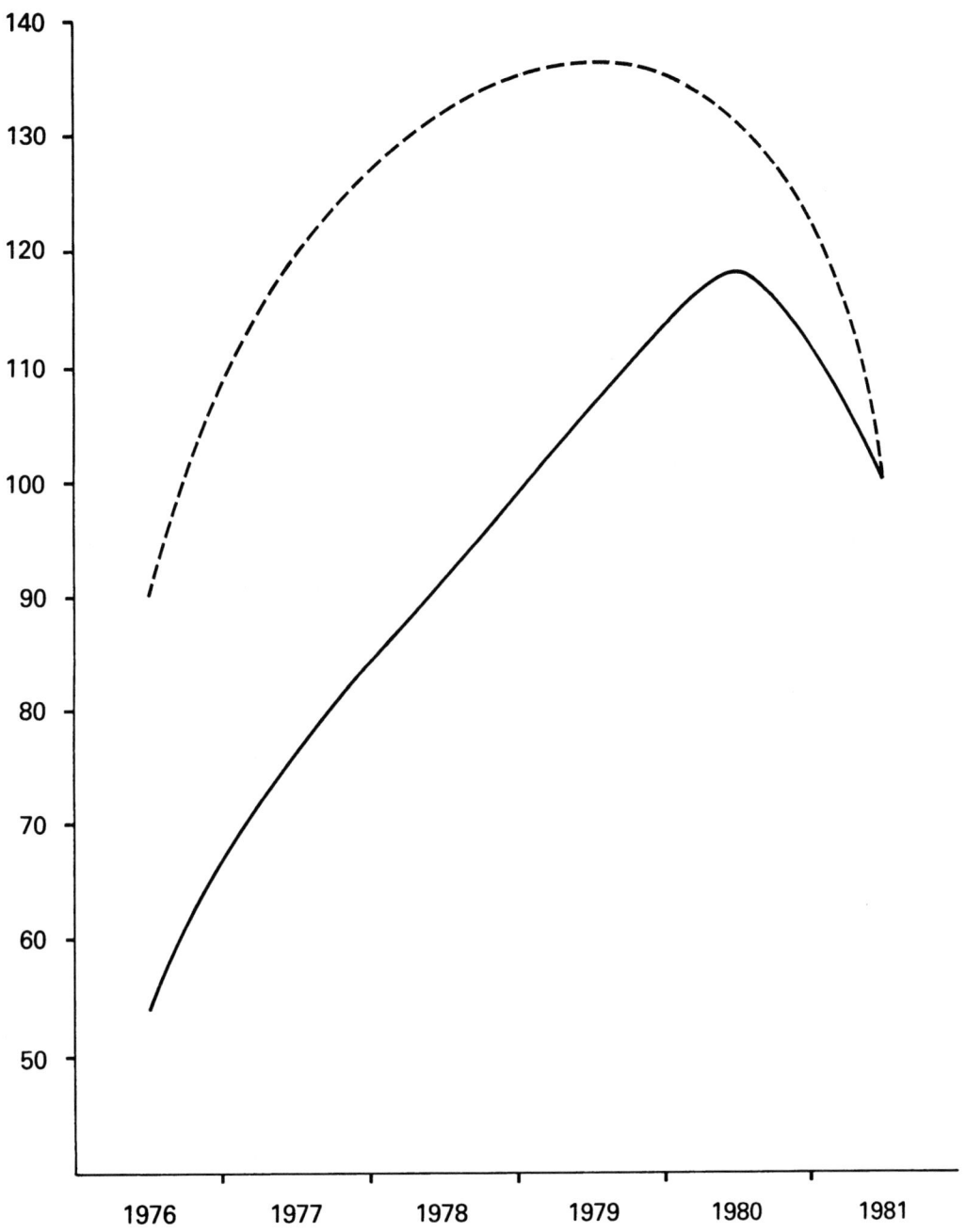

ITC Buffer Price Ranges

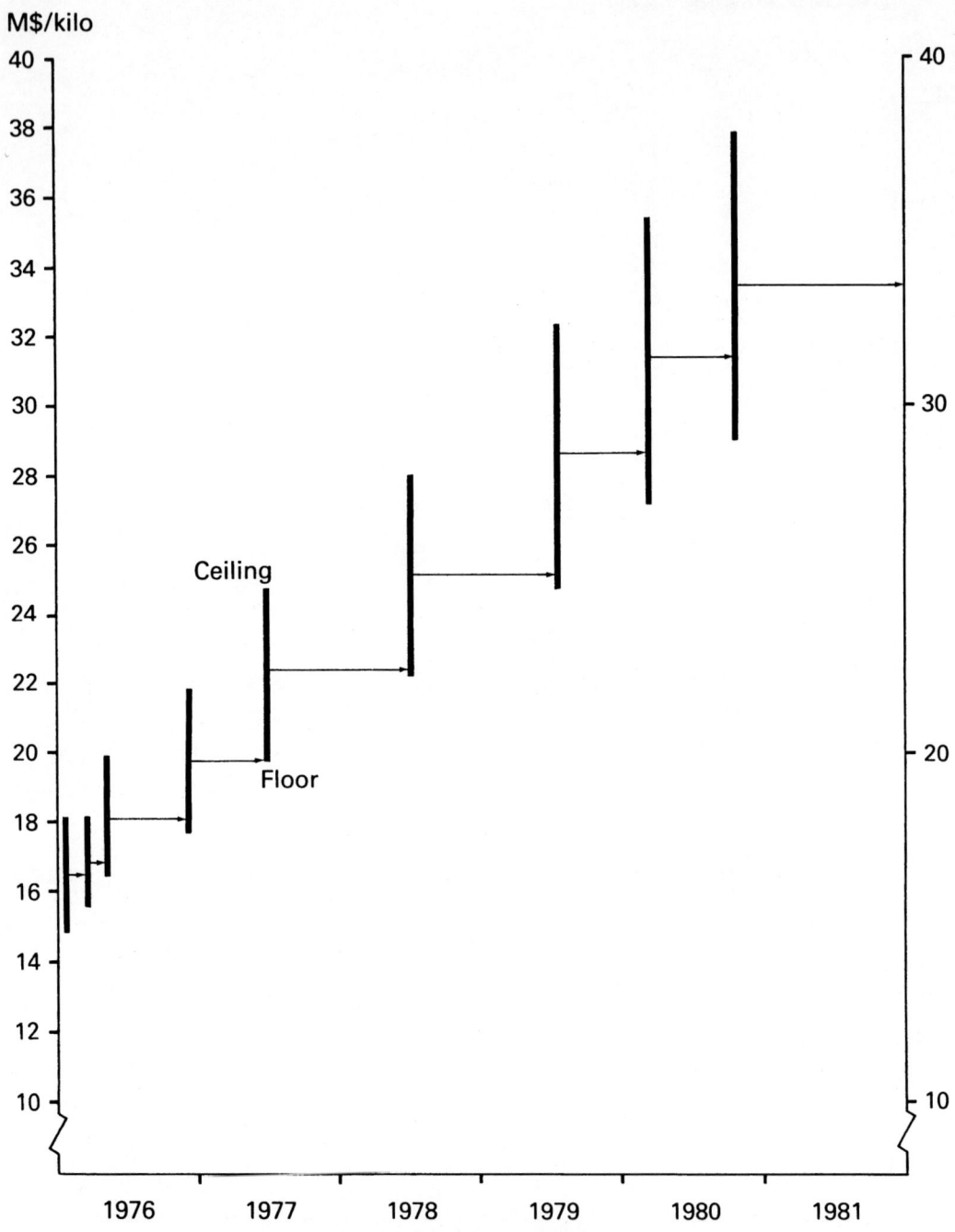

SUPPLY AND DEMAND BY MAIN MARKET AREA

	UK	EC (Ten)	Japan	USA
Production (1979/80 Averages) tonnes				
Tin in concentrates	2701	2701	604	96
Primary metal	6927	12417.5	1285.5	4214
Recycled Tin Metal	1800	3713.5	-	1806
Net Imports (1979/80 Averages) tonnes				
Tin in concentrates	7649.5	12408	1.5	2684
Tin Metal	6452	42516	30773	47168
Source of Net Imports (%)				
Tin in concentrates and Metal				
Australia	5	2		
Canada				1
European Community	10			1
S Africa	1	3		1
United States	1			
China		2	1	
Argentina	6	1		
Bolivia	41	17		15
Brazil				3
Indonesia		20	21	12
Malaysia	7	22	59	39
Nigeria	15	4		
Peru	2			..
Rwanda		1		
Singapore		1		2
Thailand		18	20	23
Zaire		4		
Others	12	5		3
Net Exports (1979/80 Averages) tonnes				
Tin in concentrates	1599.5	150.5	-	-
Tin Metal	6627.5	6885	64	3855.5

	UK	EC (Ten)	Japan	USA
Consumption (1979/80 Averages) tonnes				
Primary metal	8769.5	46796.5	31049	46919
Recycled metal	2800	4913.5	-	2200
Total metal	11569.5	51710	31049	49119
Import Dependence				
Imports as % of consumption	100	100	99	100
Imports as % of consumption and net exports	71	93	99	94
Share of World Consumption %				
Western world	7	30	18	28
Total world	5	23	14	21
Consumption Growth % p.a.				
1960s	-1.6	-0.1	6.9	1.3
1970s	-4.9	-1.7	1.9	-1.6

****

TITANIUM

WORLD RESERVES OF ILMENITE
(million tonnes of contained titanium and % of total)

Developed			Less Developed			Centrally Planned			Total
Australia	16	(8.2)	Brazil	1	(0.5)	USSR	4	(2.0)	
Canada	44	(22.4)	Egypt	1	(0.5)				
Finland	3	(1.5)	India	45	(23.0)				
Norway	36	(18.4)	Sri Lanka	1	(0.5)				
S Africa	30	(15.3)							
USA	15	(7.7)							
Totals	144	(73.5)		48	(24.5)		4	(2.0)	196

Identified world resources total some 540 million tonnes of contained titanium.

WORLD RESERVES OF RUTILE
(million tonnes of contained titanium and % of total)

Developed			Less Developed			Centrally Planned			Total
Australia	5.4	(7.8)	Brazil	54.4	(78.5)	USSR	1.5	(2.2)	
Italy	2.4	(3.5)	Sierra Leone	1.6	(2.3)				
S Africa	2.9	(4.2)	Sri Lanka	0.2	(0.3)				
USA	0.9	(1.3)							
Totals	11.6	(16.7)		56.2	(81.1)		1.5	(2.2)	69.3

Identified world resources are approximately 150 million tonnes of contained titanium.

WORLD PRODUCTION OF TITANIUM MINERALS
('000 tonnes of concentrates Averages)

Developed		Less Developed		Centrally Planned		Total
Ilmenite						
Australia (a)	1254	Brazil	16	USSR	415	
Finland	135	India	162			
Norway	688	Malaysia	180			
Portugal	..	Sri Lanka	57			
USA	539					
Rutile						
Australia	286	Brazil	..	USSR	9	
S Africa	45	India	6			
USA	n/a	Sierra Leone	30			
		Sri Lanka	15			
Slags						
Canada	676					
Japan	..					
S Africa	315					

(a) Includes leucoxene

('000 tonnes of contained TiO_2 and % of total 1979/80 Averages)

Australia	1000	(32.1)	Brazil	10	(0.3)	USSR	240	(7.7)	
Canada	480	(15.4)	India	95	(3.0)				
Finland	80	(2.6)	Malaysia	105	(3.4)				
Norway	400	(12.8)	Sierra Leone	28	(0.9)				
S Africa	310	(9.9)	Sri Lanka	47	(1.5)				
USA	325	(10.4)							
Totals	**2595**	**(83.2)**		**285**	**(9.1)**		**240**	**(7.7)**	**3120**

WORLD PRODUCTION OF TITANIUM METAL
(tonnes of sponge and % of total 1979/80)

Developed			Less Developed	Centrally Planned		Total
Japan	16.3	(20.1)		USSR	40 (49.3)	
UK	2.4	(2.9)				
USA	c.22.5	(27.7)				
Totals	41.2	(50.7)			40 (49.3)	81.2

RESERVE/PRODUCTION RATIOS

Static reserve life (years)	88 (ilmenite and rutile combined)
Ratio of identified resources to cumulative demand 1981-2000 (ilmenite and rutile combined)	14 : 1

CONSUMPTION

Titanium pigments

	1979/80 Averages '000 tonnes	% p.a. Growth rates 1970s
European Community	700 to 750	stagnating
Japan	163	4.0
United States	716	1.3

Titanium Sponge

	1979/80 Averages '000 tonnes	% p.a. Growth rates 1970s
European Community	over 10	n/a
Japan	10	12.9
United States	23	6.2

END USE PATTERNS 1980 (USA) %

Ore (Ilmenite and Rutile)

Pigment manufacture (titanium dioxide)	92
Sponge production, welding rod coats and carbides, ceramic and glass formulations	8

Metal

Aircraft and aerospace	60
Chemical processing, power generation, marine and ordnance	20
Steel and other alloys	20

VALUE OF ANNUAL PRODUCTION

$1.2 billion ($TiO_2$ content at average 1981 prices)

SUBSTITUTES

There are no effective substitutes in aircraft and aerospace applications. Nickel steels, stainless steel and some superalloy metals can sometimes be used to a limited extent in industrial uses.

Tungsten carbide competes with titanium carbide as a cutting surface in machine tools.

Materials such as zinc oxide, silica and alumina can be used in place of titanium dioxide pigment but only with loss of quality.

Synthetic rutile can be susbtituted for natural rutile.

TECHNICAL POSSIBILITIES

Environmental problems mean that plants are likely to use chloride technology in the future. Increasing amounts of synthetic rutile (made from ilmenite) are likely to be used as feed.

Possibility of using ilmenite in drilling mud applications.

Improvement in titanium forming processing to cut costs.

PRICES

	1976	1977	1978	1979	1980	1981
Ore Rutile conc 95-97% TiO_2 bagged						
A$/tonne	290	216.2	185.9	271.8	335	310
Ore Ilmenite bulk conc min 54% TiO_2						
A$/tonne	15	15	16.9	18	19.5	23.5
Sponge: US $/lb	2.7	2.72	3.08	3.84	5.5	7.47

Ore prices mainly fixed on contract. Metal prices are usually quoted by mills. Discounting common.

MARKETING ARRANGEMENTS

Moderately integrated industry with limited number of countries producing ore. Large chemical and industrial companies, e.g. Du Pont, dominate pigment production, and often have captive ore supplies.

Index Numbers 1981 = 100

The solid line gives prices in money terms and the dotted line gives prices in 'real' 1981 terms

TITANIUM
Rutile Ore

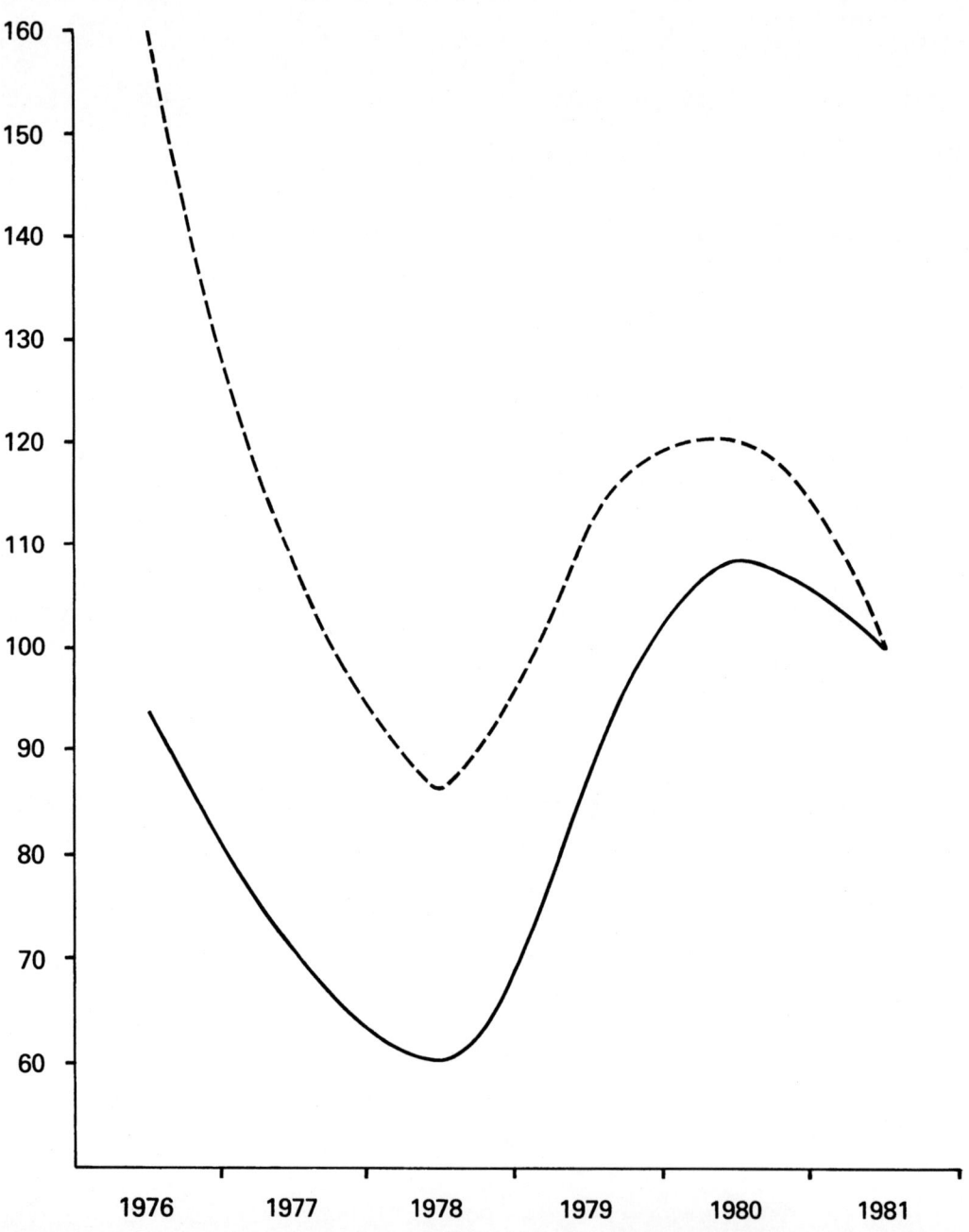

Index Numbers 1981 = 100

The solid line gives prices in money terms and the dotted line gives prices in 'real' 1981 terms

TITANIUM
US, Sponge

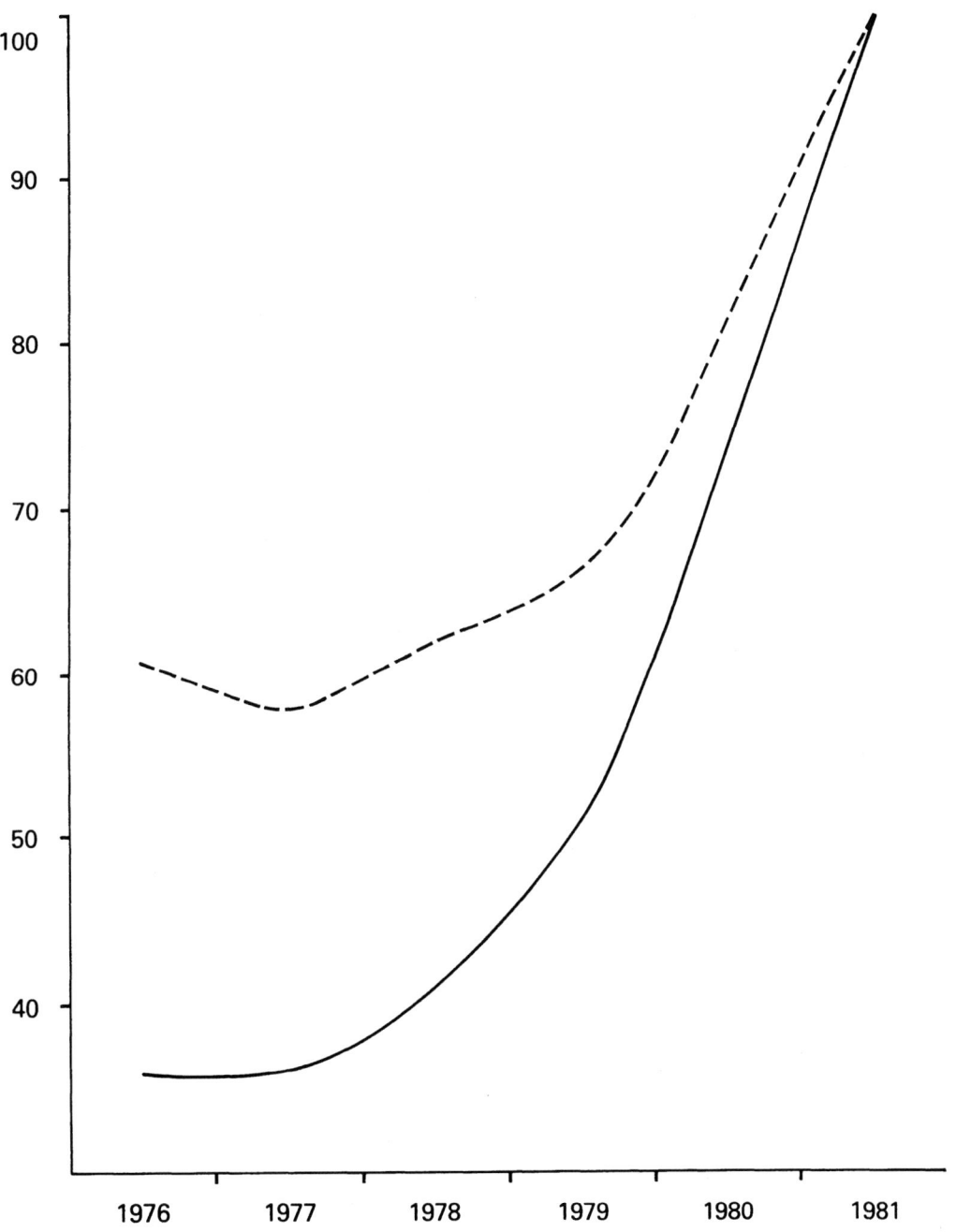

SUPPLY AND DEMAND BY MAIN MARKET AREA

	UK	EC (Ten)	Japan	USA
Production (1979/80 Averages) '000 tonnes				
Raw Materials				
Ilmenite concentrate	-	-	-	539 (TiO$_2$ 325)
Rutile concentrate	-	-	-	n/a
Titanium slags	-	-	-	-
Titanium dioxide pigments	190	n/a	179	660 (TiO$_2$ 615)
Sponge metal	2.4	2.4	16.28	c.22.5
Ingots (inc. scrap and imported sponge)	n/a	n/a	n/a	35.9
Net Imports (1979/80 Averages) '000 tonnes				
Raw Materials				
Ilmenite concentrate	282	957.6)	414.8	245.9 (TiO$_2$ 165.3)
Rutile concentrate	65.7	166.7)		256.3 (TiO$_2$ 240.9)
Titanium slag	c.39	n/a	114.3	138.9 (TiO$_2$ 102.7)
Titanium oxides (inc. pigments)	31.6	20.4	4.0	91.9
Sponge and metal (inc. wrought)	c.8.7 (inc. scrap)	over 9	-	9.1

Source of Net Imports (%)

Ilmenite, rutile and slag (TiO$_2$ content)   (exc. slag)

	UK	EC (Ten)	Japan	USA
Australia	69	39	17	65
Canada	6	18	20	15
Norway	17	37		
S Africa	6	3	8	8

	UK	EC (Ten)	Japan	USA
India			13	4
Malaysia			35	
Sierra Leone				4
Sri Lanka		1	7	
Others	1	2		4

Titanium Oxides — UK: n/a, EC (Ten): n/a

Australia				6
Canada				15
European Community			67	56
Finland			4	5
Japan				5
Norway				3
S Africa				1
Spain				8
United States			6	-
Others				1

Metal (inc. scrap) — UK: n/a, EC (Ten): n/a, Japan: -

Canada				8
European Community				13
Japan				36
China				8
USSR				28
Others				7

Net Exports (1979/80 Averages)
'000 tonnes

Raw Materials

	UK	EC (Ten)	Japan	USA
Ilmenite concentrate)			(-	n/a
Rutile concentrate)	1.1	12.7	(-	12.6
Titanium slag	-	-	-	-
Titanium dioxide pigments etc.	101.4	59.1	36.9	44.1
Sponge and metal	c.6.8 (inc. scrap)	c.1	6.48	7.9

	UK	EC (Ten)	Japan	USA
Consumption (1979/80 Averages) '000 tonnes				
Raw Materials				
Ilmenite	n/a	n/a	n/a	745.8
				(TiO$_2$ 454.8)
Rutile	n/a	n/a	n/a	260.9
				(TiO$_2$ 244)
Slag	n/a	n/a	n/a	148
				(TiO$_2$ 109)
Titanium dioxide pigments	c.120	700 to 750	163.3	715.6
				(TiO$_2$ 664)
Sponge metal	c.3.3	over 10	9.8	23
Scrap	n/a	n/a	n/a	13.3
Import Dependence (all forms)				
Imports as % of consumption	100	100	100	87
Imports as % of consumption and net exports	100	100	100	80
Share of World Consumption %				
Sponge, Metal, Pigments, Total world	4	over 12	12	28
Consumption Growth % p.a.				
1970s				
Metal	n/a	n/a	12.9	6.2
Pigments etc.	-1.5	n/a	4.0	1.3
	(1974 to 80)			
Total all forms	-1.3	flat	4.3	1.5

****

TUNGSTEN

WORLD RESERVES
('000 tonnes of metal and % of total)

Developed			Less Developed			Centrally Planned			Total
Australia	109	(4.2)	Bolivia	39	(1.5)	China	1360	(52.4)	
Austria	18	(0.7)	Brazil	18	(0.7)	N Korea	109	(4.2)	
Canada	270	(10.4)	Burma	32	(1.2)	USSR	213	(8.2)	
France	16	(0.6)	S Korea	82	(3.2)	Others	5	(0.2)	
Portugal	24	(0.9)	Malaysia	14	(0.5)				
Turkey	77	(3.0)	Mexico	20	(0.8)				
USA	125	(4.8)	Thailand	18	(0.7)				
Others	28	(1.1)	Zimbabwe	5	(0.2)				
			Others	13	(0.5)				
Totals	667	(25.7)		241	(9.3)		1687	(65.0)	2595

The world's identified resources total some 6¾ million tonnes.

WORLD MINE PRODUCTION
('000 tonnes of contained tungsten and % of total 1979/80 Averages)

Developed			Less Developed			Centrally Planned			Total
Austria	1.50	(3.0)	Argentina	0.06	(0.1)	China	12.5	(24.9)	
Australia	3.26	(6.5)	Bolivia	3.24	(6.5)	Czecho-slovakia	0.08	(0.2)	
Canada	3.13	(6.2)	Brazil	1.19	(2.4)	N Korea	2.18	(4.3)	
France	0.60	(1.2)	Burma	0.72	(1.4)	USSR	8.71	(17.4)	
Japan	0.69	(1.4)	S Korea	2.73	(5.4)				
Portugal	1.54	(3.1)	Malaysia	0.06	(0.1)				
Spain	0.28	(0.6)	Mexico	0.26	(0.5)				
Sweden	0.32	(0.6)	Namibia	0.16	(0.3)				
Turkey	1.0	(2.0)	Peru	0.56	(1.1)				
UK	0.07	(0.1)	Rwanda	0.48	(1.0)				
USA	2.88	(5.7)	Thailand	1.72	(3.4)				
			Uganda	0.05	(0.1)				
			Zaire	0.09	(0.2)				
			Zimbabwe	0.11	(0.2)				
Totals	15.27	(30.4)		11.43	(22.8)		23.47	(46.8)	50.17

RESERVE/PRODUCTION RATIOS

Static reserve life (years)   52
Ratio of identified resources
to cumulative demand 1981-2000   5.7 : 1

CONSUMPTION OF TUNGSTEN CONCENTRATES

	1979/80 Averages tonnes	% p.a. Growth rates 1970s
European Community	4890	-6
Japan	2245	-4.9
United States	9530 (a)	1.6
Other Countries	6825	4.1
Total Western world	23490	1.1
Total world	39110	0.6

(a) Some 800 to 1500 tonnes of scrap are also used

These statistics derived from the UNCTAD Committee on Tungsten show the immediate consumption of tungsten concentrates, but not necessarily the final destination of the subsequent products.

END USE PATTERNS 1980 (USA) %

Metal working and construction machinery	78
Transport	9
Lamps and lighting	6
Electrical	4
Other	3

VALUE OF CONTAINED METAL IN ANNUAL PRODUCTION

$65 million (at average 1981 prices)

SUBSTITUTES

Titanium, tantalum and columbium carbides can be used in some wear-resisting applications.

Molybdenum tool steels and tungsten tool steels are interchangeable.

In some cutting tool applications, bulk ceramics are an alternative.

TECHNICAL POSSIBILITIES

Further development of new metal-shaping methods e.g. laser.

Increased use of tungsten scrap.

The life of cemented carbide cutting tool inserts has been increased recently by use of coatings. This trend is continuing.

PRICES

	1976	1977	1978	1979	1980	1981
Ore min 65% WO_3 cif Europe						
$/mtu WO_3 (range)	85-148	142-186	128-173	115-151	131-154	120-155

Prices traditionally highly volatile. Consumers buy direct from producers or through traders, often by reference to Metal Bulletin price quotes which usually represent small spot lots (or by reference to Tungsten Users Index which is based on samples of purchases). Long term contractual prices usually lower.

MARKETING ARRANGEMENTS

Market influenced by two extraneous factors - exports from China and sales from the US stockpile. Major producers, with exception of Canada, meet under auspices of Primary Tungsten Association. More important is the UNCTAD Tungsten Committee which collects data and attempts, unsuccessfully so far, to reach agreement on market stabilisation measures. Increasing interest, especially in US, in integrated production through construction of tungsten conversion plants.

Index Numbers 1981 = 100

The solid line gives prices in money terms and the dotted line gives prices in 'real' 1981 terms

TUNGSTEN
65% WO₃. cif Europe
(Lower limit)

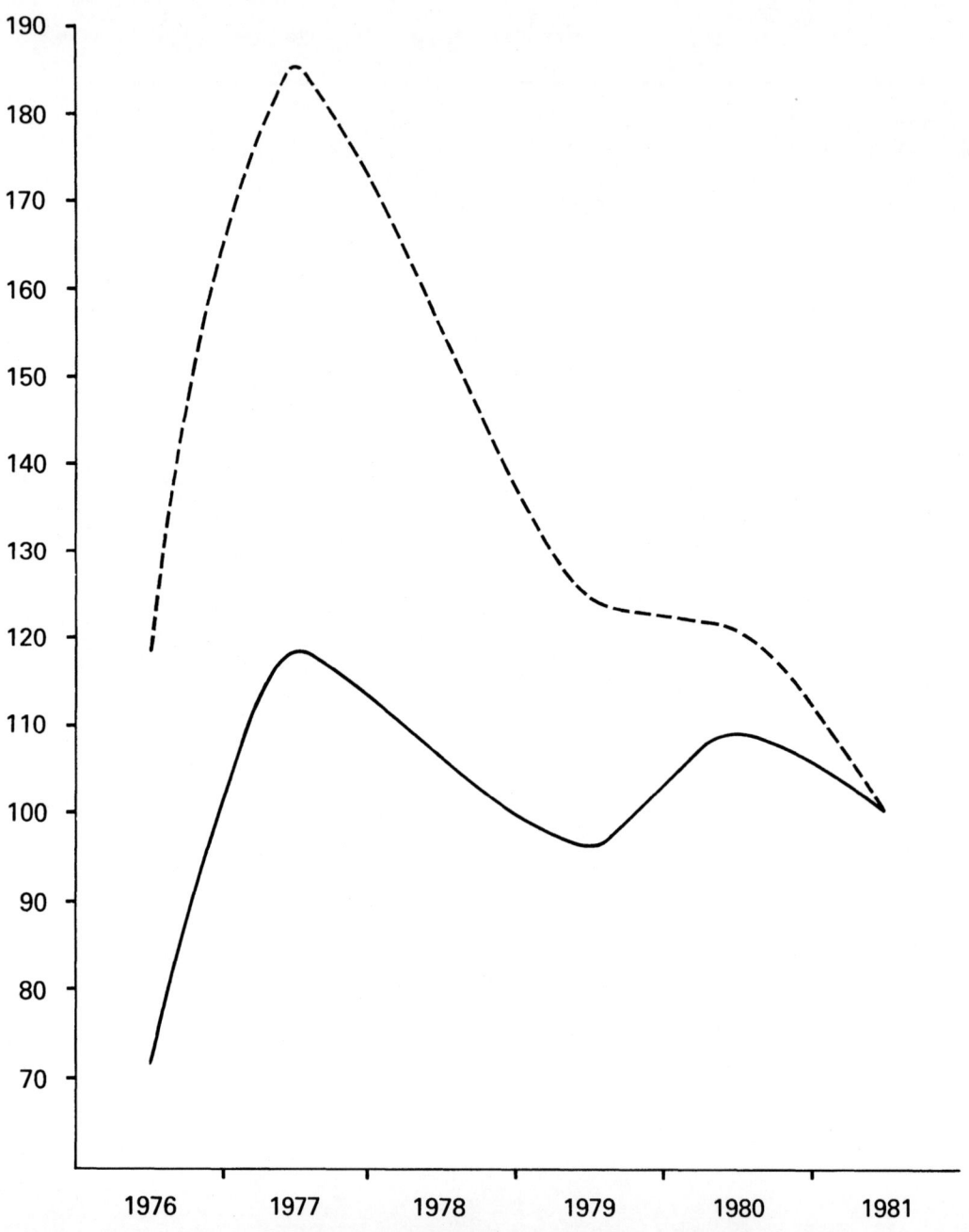

SUPPLY AND DEMAND BY MAIN MARKET AREA

		UK	EC (Ten)	Japan	USA
Production (1979/80 Averages) tonnes					
Mine production	(W content)	67	662	693	2883 (a)
Ammonium paratungstate	(W content)	n/a	n/a	n/a	7860

(a) There have been US Stockpile releases in recent years (1979/80 average 2027 tonnes)

Net Imports (1979/80 Averages) tonnes		UK	EC (Ten)	Japan	USA
Concentrates	(W content)	1534.5	c.4060	1840	5154
Ferro tungsten	(W content)	170	350	c.268	230
Ammonium paratungstate	(W content)	n/a	n/a	n/a	200
Tungsten metal		271.5	891.5	160	344
Tungsten carbide	(W content)	374 (gross)	n/a	n/a	243

Source of Net Imports (%)

Ores and concentrates

	UK	EC (Ten)	Japan	USA
Australia	2	12	13	3
Canada		8	17	27
European Community	39			
Portugal	26	14	7	3
Spain	2	3		
United States	4		7	
China	5	32	6	14
Bolivia			7	25
Burma		2		
Mexico				5
Peru		1	9	6
Rwanda		3		
S Korea		4	23	
Thailand	6	9	7	10
Others	16	12	4	7

Net Exports (1979/80 Averages)
 tonnes

Concentrate	(W content)	71	c.1013	–	898
Metal and powder	(W content)	280 (gross)	297	–	c.1000
Tungsten carbide powder	(W content)	469	n/a	–	642
Ferro tungsten	(W content)	505	630	–	n/a

Consumption (1979/80 Averages)
 tonnes

Concentrates	(W content)	1512	4890	2245	9530
Apparent Total	(W content)	c.1000	c.5250 to 5500	2675	10337 (concentrates, scrap & metal)

Import Dependence (all forms)

Imports as % of consumption	100	100	85	60 (inc. scrap)
Imports as % of consumption and net exports	100	78	85	48 (inc. scrap)

Share of World Consumption of Concentrates %

Western world	6	21	10	41
Total world	4	13	6	24

Consumption Growth % p.a.

1970s	-8.3	-6	-4.9	1.6

****

URANIUM

WESTERN WORLD REASONABLY ASSURED RESOURCES
('000 tonnes U and % of total)

Developed			Less Developed			Total
Australia	317	(13.8)	Algeria	26	(1.1)	
Austria	0.3	-	Argentina	30.3	(1.3)	
Canada	258	(11.2)	Brazil	119.1	(5.2)	
Denmark	27	(1.2)	Central African Republic	18	(0.8)	
Finland	3.4	(0.1)	Chile	..	..	
France	74.9	(3.3)	Gabon	21.6	(0.9)	
W Germany	5	(0.2)	India	32	(1.4)	
Greece	5.4	(0.2)	S Korea	11	(0.5)	
Italy	2.4	(0.1)	Mexico	2.9	(0.1)	
Japan	7.7	(0.3)	Namibia	135	(5.9)	
Portugal	8.2	(0.4)	Niger	160	(7.0)	
S Africa	356	(15.5)	Somalia	6.6	(0.3)	
Spain	16.4	(0.7)	Zaire	1.8	(0.1)	
Sweden	38	(1.7)				
Turkey	4.6	(0.2)				
USA	605	(26.4)				
Totals	1729.3	(75.4)		564.3	(24.6)	2293.6

The table includes estimates of reasonably assured resources available at 1st January 1981. Some 76% of the total is available at an estimated forward cost of (not price) $80/Kg U or less, and the balance at a forward cost of $80 to 130 Kg/U. Estimated additional resources in the Western world, available at a forward cost under $130/Kg of U amount to 2.72 million tonnes of U.

WESTERN WORLD MINE PRODUCTION
(tonnes U and % of total 1979/80 Averages)

Developed			Less Developed			Total
Australia	1133	(2.8)	Argentina	160.5	(0.4)	
Canada	6985	(17.0)	Gabon	1066.5	(2.6)	
France	2498	(6.1)	Namibia	3941	(9.6)	
W Germany	30	(0.1)	Niger	3860	(9.4)	
Japan	3.5	(..)				
Portugal	98	(0.2)				
S Africa	5471.5	(13.3)				
Spain	190	(0.5)				
USA	15600	(38.0)				
Totals	32009	(78.0)		9028	(22.0)	41037

Note:- There is presently a limited amount of reprocessing of spent reactor fuel which supplements mine production.

RESERVE/PRODUCTION RATIOS

Static reserve life : Western world (years) : 56
Ratio of resources to cumulative demand
1981 to 2000 (includes reasonably assured and
estimated additional resources : 3.9 : 1

(based on an OECD medium projection for a
LWR once through strategy. The ratio would
not differ significantly for other strategies).

CONSUMPTION

The table shows average Western world reactor consumption in 1979/80, based on estimated reactor usage. Stockpiling is ignored.

	1979/80 Averages tonnes (U)	% p.a. Growth rates 1970s
European Community (Ten)	8419	14.3
Japan	2083	25.3
United States	8036	10.6
Others	5195	20.5
Total Western world	23733	14.5

Source: Nuexco No. 168

The future growth in demand depends on the integrity of nuclear reactor programmes. Because of delays to reactor construction, and stockpiling purchases of uranium have substantially exceeded reactor usage.

END USE PATTERNS

The only uses for natural uranium are for military purposes and for civil nuclear power. There is no published breakdown of demand between the two, although nearly all recent output goes into the latter.

VALUE OF ANNUAL PRODUCTION

$3.2 billion at an average value of $30/lb

SUBSTITUTES

Nuclear power directly competes with other forms of electricity generation. Once a nuclear station is built, however, there is no substitute for uranium based fuel. Reprocessing of spent fuel allows some limited substitution for mined output in the longer term, but is constrained in the short term by a lack of reprocessing facilities and waste storage.

TECHNICAL POSSIBILITES

Improvements in the technology of existing reactor types would allow a modest (10-15%) saving in uranium requirements. Longer term, the fast breeder reactor, now at the prototype stage, would allow a substantial (60 fold) reduction in uranium usage per unit of power. Its development has been delayed and it will probably not be a commercial proposition until well into the 21st. century. Work on different types of reactor to the light water reactor has been inhibited by developments in the energy and uranium markets but several processes have been developed for processing and enrichment.

PRICES

	1976	1977	1978	1979	1980	1981
Nuexco exchange value $/lb U_3O_8	39.7	42.2	43.2	42.6	31.8	24.2
US DOE Average Contract Price $/lb U_3O_8	16.1	19.8	21.6	23.9	26	32.2 (Buyers)

Index Numbers 1981 = 100

The solid line gives prices in money terms and the dotted line gives prices in 'real' 1981 terms

URANIUM
Nuexco exchange value

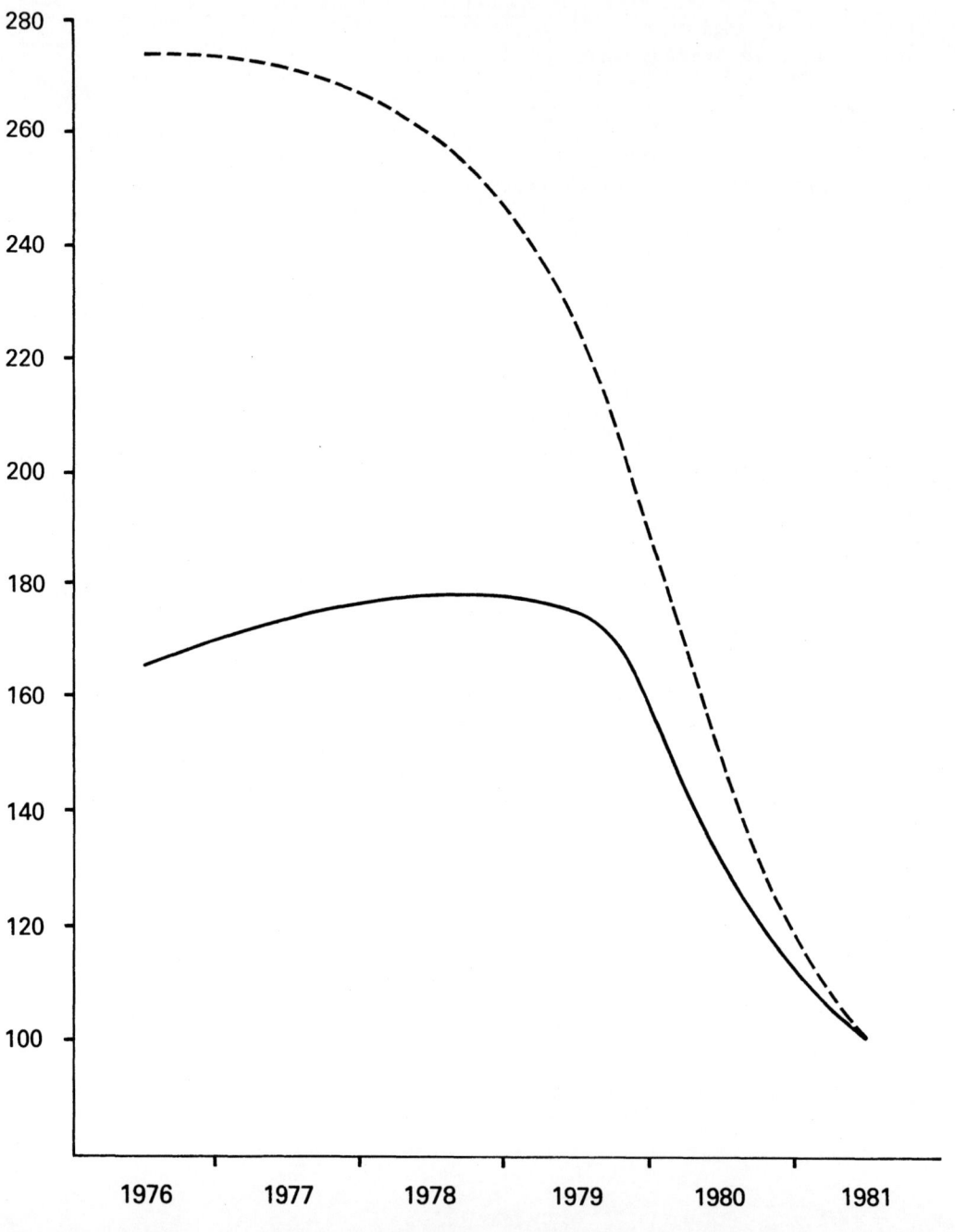

MARKETING ARRANGEMENTS

Uranium is a minor but very important element of the cost of nuclear power. Most sales are under long term contracts directly negotiated between mines and utilities. The intermediate processing from uranium concentrate to fuel rods is carried out under tolling arrangements. Because of its sensitive nature, uranium exports are heavily controlled by host governments in nearly all respects. There is a small spot market organised by uranium brokers in which utilities can dispose of excess supplies. Spot prices are only one influence on prices fixed under long term contracts. Production costs are also important. There is a limited number of mines, which are controlled by a mixture of government agencies, utilities, oil companies and mining companies. Most producers have large interests outside uranium mining.

SUPPLY AND DEMAND BY MAIN MARKET AREA

Data on foreign trade compatible with production and consumption statistics are not published. On the basis of mine production and reactor usage Japan and the United Kingdom are completely reliant on imports or accumulated inventories. Production in the European Community equalled 30% of its reactor usage, whilst in the United States average 1979/80 consumption greatly exceeded final consumption in nuclear reactors. The processing of natural uranium (conversion, enrichment and fuel fabrication) is a constraint on commercial freedom. Supplying countries also impose various restrictions on trade in, and the uses of their exports.

****

VANADIUM

WORLD RESERVES
('000 tonnes of contained vanadium and % of total)

Developed			Less Developed			Centrally Planned		Total
Australia	180	(1.1)	Chile	136	(0.9)	USSR	7250 (45.9)	
Finland	127	(0.8)	India	90	(0.6)			
Norway	18	(0.1)	Venezuela	90	(0.6)			
S Africa	7800	(49.4)						
USA	104	(0.7)						
Totals	8229	(52.1)		316	(2.0)		7250 (45.9)	15795

The world's identified resources amount to 56 million tonnes. Most of the resources are in titaniferous-magnetites from which vanadium would be produced as a by-product of iron, as they are in crude petroleum and tar sands. In all these cases extraction depends on economic recovery of the main product.

WORLD MINE PRODUCTION
('000 tonnes of contained vanadium and % of total 1979/80 Averages)

Developed			Less Developed			Centrally Planned		Total
Australia (a)	0.6	(1.7)	Chile	0.4	(1.1)	China	4 (11.2)	
Finland	2.8	(7.9)				USSR	10 (28.1)	
Norway	0.6	(1.7)						
S Africa	12.5	(35.1)						
USA	4.7	(13.2)						
Totals	21.2	(59.6)		0.4	(1.1)		14 (39.3)	35.6

(a) 1980 only.

RESERVE/PRODUCTION RATIOS

Static reserve life (years) 444
Ratio of identified resources
to cumulative demand 1981-2000   57 : 1

CONSUMPTION

	1979/80 Averages (tonnes V)	% p.a. Growth rates 1970s
European Community (Ten)	c.6000	n/a probably fell
Japan	2766	8.7
United States	6640	-0.3

END USE PATTERNS 1980 (USA) %

Transport	35
Construction	27
Machinery	25
Chemicals	4
Other	9

VALUE OF ANNUAL PRODUCTION

$300-350 million (at average 1981 prices)

SUBSTITUTES

Heat-treated carbon steels and various other alloy steels can be substituted for steels containing vanadium.

Platinum can be used in some catalytic processes but at higher cost.

To some extent, columbium, molybdenum, manganese, titanium and tungsten are interchangeable in many applications.

TECHNICAL POSSIBILITIES

Possible recovery from low grade dolomitic shales and sandstones and extractions from tar sands and oil shales.

Potential use in superconductors and batteries.

PRICES

	1976	1977	1978	1979	1980	1981
US Pentoxide/Chemical $/lb V_2O_5	3.08	3.15	3.23	3.39	3.86	4.20
US, Pentoxide/ Metallurgical (fused) $/lb V_2O_5	2.89	3.05	3.34	3.40	3.55	3.52
Ferrovanadium US Producer 80% V. $/lb V	6.23	6.05	6.25	6.85	7.75	8.5

Mainly producer pricing.

MARKETING ARRANGEMENTS

Mainly produced as by-product or co-product of other metals, especially iron and uranium. S Africa, especially Highveld Steel and Vanadium Corp, and USSR are world's major producers with China becoming increasingly important. Trend towards vertical integration through ferrovanadium production facilities.

Index Numbers 1981 = 100

The solid line gives prices in money terms and the dotted line gives prices in 'real' 1981 terms

VANADIUM
US, Pentoxide/Chemical

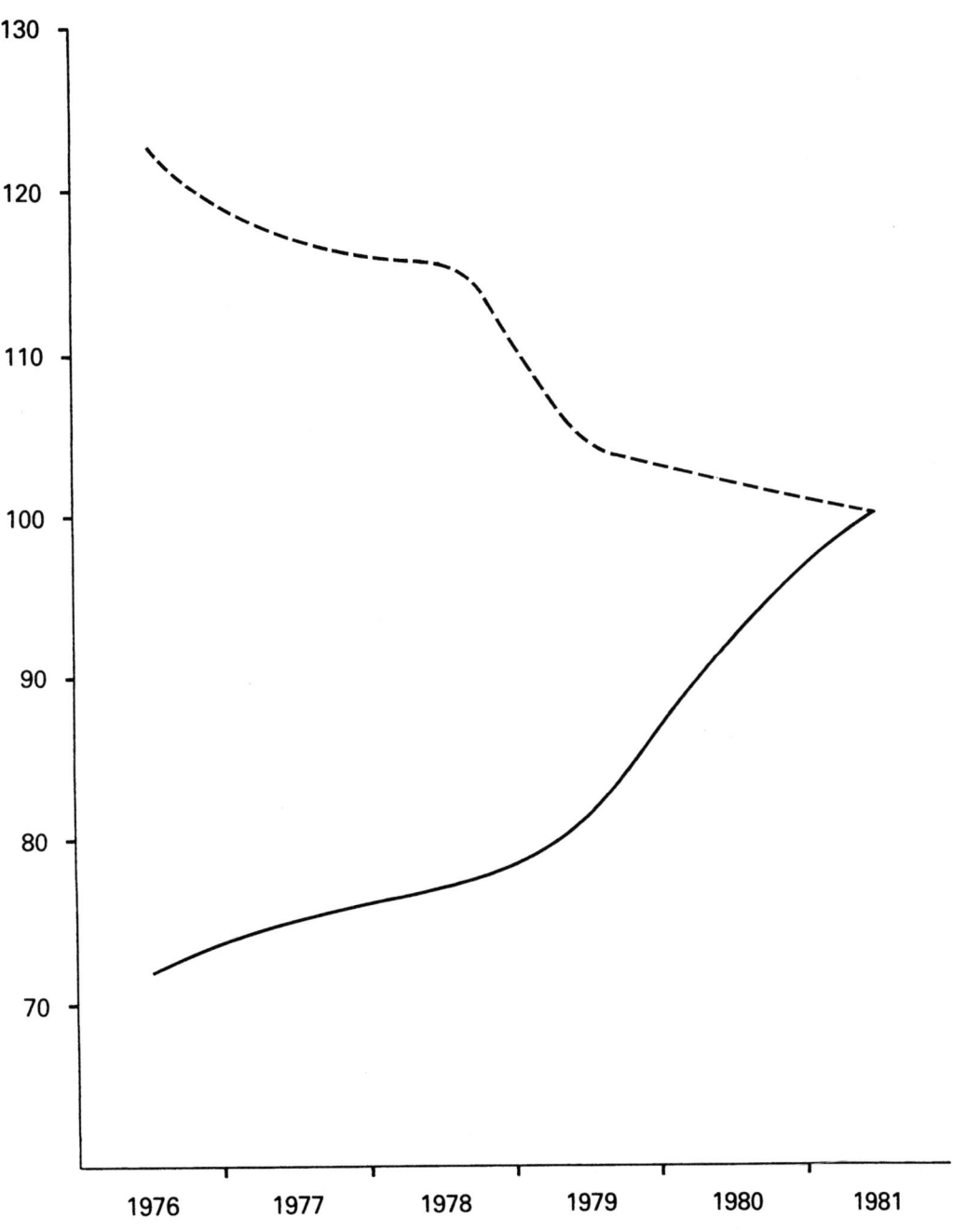

SUPPLY AND DEMAND BY MAIN MARKET AREA

	UK	EC (Ten)	Japan	USA
Production (1979/80 Averages) tonnes				
Ores and concentrates : Mine (V content)	-	-	-	4684
Mill (V content)	-	-	-	5109
Ferro vanadium	-	c.7750 (V cont. c.5540)	4222 (V cont. c.2955)	8910 (V cont. 5345)
Net Imports (1979/80 Averages) tonnes				
Ores, slags, residues	567.5 (gross)	46044 (gross)	-	1918 (V cont.)
Vanadium oxides	931 (V. cont. c.520)	3549 (V cont. c.1990)	4017.5 (V cont. c.2250)	800 (V cont.)
Ferro vanadium	513.5 (V. cont. c.360)	1510.5 (V cont. c.1060)	786.5 (V cont. c.550)	383 (V cont.)
Vanadium metal	97	70		
Source of Net Imports (%)				
Vanadium pentoxide				
Austria		16		
European Community	6		6	
Finland	91	38		48
S Africa	3	34	89	51
United States		2	3	
China		9	1	
Others		1	1	1
Ferro vanadium				
Austria	57	63	24	1
Canada			4	58
European Community	17		36	31
Norway	16	23		
Sweden	10	5		10
United States		5	20	
Argentina		3		
Brazil			16	

	UK	EC (Ten)	Japan	USA
Net Exports (1979/80 Averages) tonnes				
Vanadium pentoxide	35 (V. cont. c.25)	158.5 (V cont. c.89)	-	344 (V cont.)
Ferro vanadium	132.5 (V. cont. c.93)	696.5 (V cont. c.488)	50 (V cont. 35)	534 (V cont.)
Vanadium compounds (inc. ores)			-	204 (V cont.)
Vanadium metal	-	33.5	-	-
Consumption (1979/80 Averages) tonnes	c.750 (V. cont.)	c.6000 (V cont.)	2766 (V cont.)	6640 (a) (V cont.)

(a) Including processing losses from low grade imports

Import Dependence

	UK	EC (Ten)	Japan	USA
Imports as % of consumption	100	100	100	47
Imports as % of consumption and net exports	100	100	100	40

Share of World Consumption %

	UK	EC (Ten)	Japan	USA
Total world (approx.)	2	17	8	19

Consumption Growth % p.a.

	UK	EC (Ten)	Japan	USA
1970s	-3 (iron & steel industry)	n/a (probably fell)	⟍ 8.7	-0.3

****

ZINC

WORLD RESERVES
(million tonnes of contained zinc and % of total)

Developed			Less Developed			Centrally Planned			Total
Australia	16	(9.9)	Brazil	9	(5.6)	China	5	(3.1)	
Canada	30	(18.5)	India	3	(1.9)	Poland	3	(1.9)	
Ireland	8	(4.9)	Iran	5	(3.1)	USSR	11	(6.8)	
Japan	5	(3.1)	Mexico	3	(1.9)	Others	4	(2.5)	
S Africa	11	(6.8)	Peru	7	(4.3)				
Spain	4	(2.5)	Other America	3	(1.9)				
USA	15	(9.3)	Other Africa	4	(2.5)				
Others	11	(6.8)	Others	5	(3.1)				
Totals	100	(61.7)		39	(24.1)		23	(14.2)	162

Identified world resources total 325 million tonnes. If hypothetical and subeconomic resources are included the total would be about 4,400 million tonnes.

WORLD MINE PRODUCTION
('000 tonnes of contained zinc and % of total 1979/80 Averages)

Developed			Less Developed			Centrally Planned			Total
Australia	511.5	(8.1)	Bolivia	50.0	(0.8)	Bulgaria	79	(1.3)	
Canada	1131.6	(18.0)	Brazil	69.3	(1.1)	China	152.5	(2.4)	
W Germany	119.0	(1.9)	S Korea	59.3	(0.9)	N Korea	132.5	(2.1)	
Greenland	89.7	(1.4)	Mexico	241.9	(3.8)	Poland	226.9	(3.6)	
Finland	56.5	(0.9)	Peru	510.8	(8.1)	Romania	49	(0.8)	
Ireland	220.4	(3.5)	Zambia	48.1	(0.8)	USSR	1010	(16.0)	
Italy	62.4	(1.0)	Zaire	70.0	(1.1)	Others	22	(0.3)	
Japan	240.8	(3.8)	Others	178.9	(2.8)				
S Africa	69.1	(1.1)							
Spain	160.7	(2.6)							
Sweden	168.3	(2.7)							
USA	330.9	(5.3)							
Yugoslavia	98.0	(1.6)							
Others	137.0	(2.2)							
Totals	3395.9	(53.9)		1228.3	(19.5)		1671.9	(26.6)	6296.1

WORLD SMELTER PRODUCTION
('000 tonnes of zinc metal and % of total 1979/80 Averages)

Developed			Less Developed			Centrally Planned			Total
Australia	303.1	(4.8)	Algeria	28.7	(0.5)	Bulgaria	90	(1.4)	
Austria	22.7	(0.4)	Argentina	31.1	(0.5)	China	157.5	(2.5)	
Belgium	250.1	(4.0)	Brazil	70.9	(1.1)	E Germany	16.5	(0.3)	
Canada	586.0	(9.3)	India	53.6	(0.9)	N Korea	112.5	(1.8)	
Finland	146.6	(2.3)	S Korea	78.3	(1.2)	Poland	212.2	(3.4)	
France	250.9	(4.0)	Mexico	153.6	(2.4)	Romania	45.8	(0.7)	
W Germany	360.4	(5.7)	Peru	66.1	(1.1)	USSR	1072.5	(17.0)	
Italy	204.8	(3.3)	Zambia	35.5	(0.6)	Vietnam	10.0	(0.2)	
Japan	764.3	(12.1)	Zaire	43.7	(0.7)				
Netherlands	161.8	(2.6)							
Norway	78.5	(1.2)							
Portugal	1.0	(..)							
S Africa	78.4	(1.2)							
Spain	168.9	(2.7)							
Turkey	15.2	(0.2)							
UK	81.7	(1.3)							
USA	447.8	(7.1)							
Yugoslavia	91.7	(1.5)							
Totals	4013.9	(63.8)		561.5	(8.9)		1717.0	(27.3)	6292.4

SECONDARY PRODUCTION : WESTERN WORLD
('000 tonnes of zinc 1979/80 Averages)

Scrap used by primary smelters (and included in primary output)	300
Remelted zinc and alloys	495
Zinc in copper and other alloys	500
Scrap as such, used by chemical etc. plants	240
Total Secondary recovery additional to smelter output	1235

RESERVE/PRODUCTION RATIOS

Static reserve life (years)	26
Ratio of identified resources to cumulative demand 1981-2000	2 : 1

CONSUMPTION

	1979/80 Averages '000 tonnes	% p.a. Growth rates 1960-70	1970-80
European Community (Ten)	1385	3.2	0.8
Japan	767	13.6	2.2
United States	904	3.3	-2.4
Others	1449	7.1	5.1
Total Western world	4505	5.1	1.4
Total world	6204	5.1	2.3

END USE PATTERNS 1980 (USA) %

Construction	40
Transport equipment	19
Electrical equipment	12
Machinery and chemicals	18
Other	11

VALUE OF CONTAINED METAL IN ANNUAL PRODUCTION

$5.5 billion (as Slab metal at average 1981 prices)

SUBSTITUTES

Aluminium, magnesium and plastics compete in some die-casting applications.

Ceramic and plastic coatings, electroplated cadmium and aluminium, and special steels compete in some galvanising applications.

Zirconium, aluminium, magnesium and titanium can replace zinc in chemicals and pigments.

Aluminium alloys, stainless steels and plastics can be used in place of brass.

TECHNICAL POSSIBILITIES

Development of new alloys e.g. superplastic alloys of zinc and aluminium.

Improvements in thin-wall zinc die-casting technology bringing new applications.

Possibility of zinc batteries after 1990.

Increased recovery of secondary zinc.

Potential for more use in the construction industry.

PRICES

	1976	1977	1978	1979	1980	1981
¢/lb						
LME Cash	32.3	26.8	26.9	33.6	34.5	38.9
US Producer	37.0	34.4	31.0	37.3	37.4	44.6
European Producer	36.1	32.4	27.5	35.8	36.1	41.3
£/tonne						
LME Cash	394.9	338.1	309.1	349.9	327.4	429.1
LME Cash Monthly	340.2-	288.6-	245.7-	296.9-	289.8-	318.1-
Average range	433.8	417.9	354.8	395.1	379.3	524.1

Most zinc metal traded at producer prices with LME a vestigial market which is, nonetheless, a major influence on producer prices, especially in present recession. Ores and concentrates purchased by smelters at producer prices less negotiated treatment charges. Mine costs influenced by by-product values. Capital costs such that new mines need over cents 50/lb.

MARKETING ARRANGEMENTS

United Nations' International Lead and Zinc Study Group an intergovernmental forum for statistical analysis and discussion of common problems. Wide geographical spread of production with over 300 mines, but much smaller number of smelters, which determine prices. Market leadership in USA, and producer pricing in Europe. Latter weakened by persistent recession, and growth of smelter production near mines in developing countries, often under state control.

Index Numbers 1981 = 100

The solid line gives prices in money terms and the dotted line gives prices in 'real' 1981 terms

ZINC
LME Cash

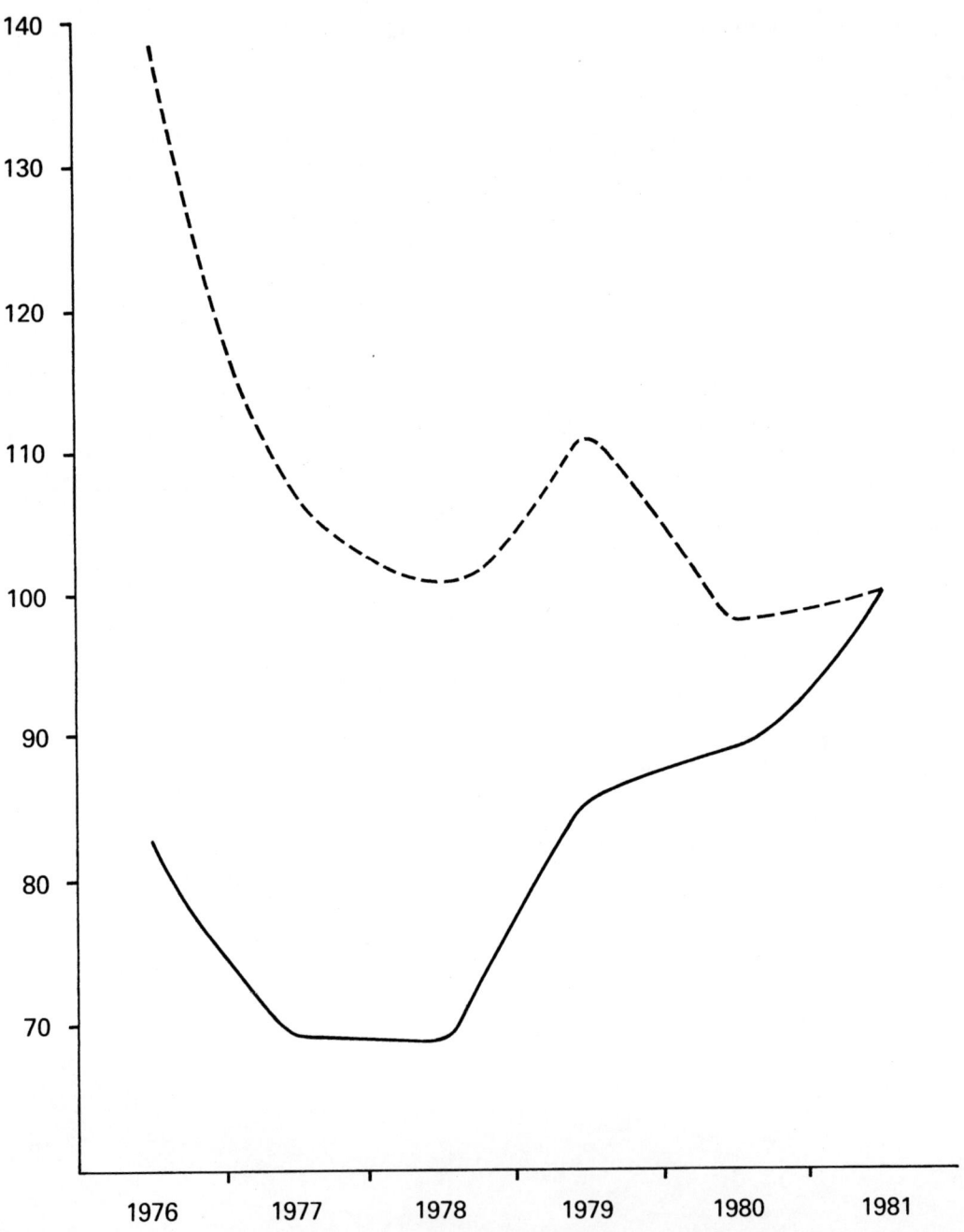

SUPPLY AND DEMAND BY MAIN MARKET AREA

	UK	EC (Ten)	Japan	USA
Production (1979/80 Averages) '000 tonnes Zn content				
Mine production	2.2	555.7 (inc. Greenland)	240.8	330.9
Smelter production	81.7	1309.6	764.3	447.8
Zinc recovered from scrap (other than zinc in primary smelters)	67.6	181.5	169	295.9
Net Imports (1979/80 Averages) '000 tonnes Zn content				
Ores and concentrates	90	814	433	177.5
Unwrought zinc metal	153.1	201.3	39.3	468.9
Source of Net Imports (%)				
Ores, concentrates and metal				
Australia	10	6	28	5
Canada	24	33	32	55
European Community	30			7
Finland	13	6		4
Japan				1
Norway	5	3		
S Africa (inc. Namibia)		2		
Spain	2	4		8
Sweden		8		
United States	1	2		
Yugoslavia				
Eastern Countries	1	3	12	
Bolivia		2	1	2
Honduras	..	1		2
Mexico		5	1	7
Morocco		1		
Peru	10	17	22	6
Philippines			2	
S Korea			1	
Zaire				1
Zambia	1	1		1
Other Countries	3	6	2	1

	UK	EC (Ten)	Japan	USA
Net Exports (1979/80 Averages) '000 tonnes Zn content				
Ores and concentrates	2.2	93.2	-	35.8
Unwrought zinc metal	19.1	120.6	39.5	0.4
Consumption (1979/80 Averages) '000 tonnes Zn content				
Primary metal	210	1385	767	904
Zinc recovered from scrap	67.6	181.5	169	295.9
Direct use of concentrates	-	-	-	71.4
Import Dependence				
Imports as % of consumption	88	65	50	51
Imports as % of consumption and net exports	81	57	48	49
Share of World Consumption %				
Western world	5	31	17	20
Total world	3	22	12	15
Consumption Growth % p.a.				
1960s	0.7	3.2	13.6	3.3
1970s	-3	0.8	2.2	-2.4

****

ZIRCONIUM

WORLD RESERVES
('000 tonnes of zirconium and % of total)

Developed			Less Developed			Centrally Planned			Total
Australia	7260	(28.9)	Brazil	910	(3.6)	USSR	2720	(10.8)	
S Africa	5440	(21.6)	India	3630	(14.4)				
USA	3630	(14.4)	Madagascar	90	(0.4)				
			Malaysia and Thailand	90	(0.4)				
			Sierra Leone	455	(1.8)				
			Sri Lanka	910	(3.6)				
Totals	16330	(65.0)		6085	(24.2)		2720	(10.8)	25135

The world's identified resources amount to some 40 million tonnes of contained zirconium.

WORLD MINE PRODUCTION
('000 tonnes of concentrates and % of total 1979/80 Averages)

Developed			Less Developed			Centrally Planned			Total
Australia	453	(60.1)	Brazil	4	(0.5)	China	c.15	(2.0)	
S Africa	81	(10.7)	India	13	(1.7)	USSR	c.75	(9.9)	
USA	c.110	(14.6)	Malaysia	1	(0.1)				
			Sri Lanka	2	(0.3)				
			Thailand	..					
Totals	c.644	(85.4)		20	(2.7)		c.90	(11.9)	c.754

The zirconium contained in this output of concentrate was roughly 35,000 tonnes.

RESERVE/PRODUCTION RATIOS

Static reserve life (years) c.70
Ratio of identified resources
to cumulative demand 1981-2000 c. 4½ : 1

CONSUMPTION

Approximate consumption of concentrates

	1979/80 Averages '000 tonnes	% p.a. Growth rates 1970s
European Community	c.190	4.2
Japan	c.175	8.4
United States	142	0.2

END USE PATTERNS 1980 (USA) %

Zircon

Foundry sands	42
Refractories	30
Ceramics	13
Abrasives	5
Other, including chemicals, and metal for nuclear applications and chemical processing equipment	10

VALUE OF ANNUAL PRODUCTION

$75 million (at average 1981 prices)

SUBSTITUTES

Chromite sand can be used in place of zircon in some foundry application. Titanium and tin compounds can replace zirconium oxide in ceramics.

A number of alternatives are available in the nuclear applications, notably stainless steel as a structural material and aluminium, columbium and vanadium for fuel containers.

Stainless steel, titanium and tantalum are substitutes in many corrosion-resistant industrial applications.

Many materials, particularly ferroalloys, compete with zirconium in ferrous metal applications.

TECHNICAL POSSIBILITIES

Production of hafnium-free zirconium by a new distillation technique, at reduced cost to conventional techniques, is at the pilot plant stage.

PRICES

	1976	1977	1978	1979	1980	1981
Zircon sand 66-67% ZrO_2						
Standard grade A$/tonne	150	109.3	70.8	58.9	56.8	84
Zirconium sponge $/lb range	5.50-7.00	5.50-9.00	9-15	9-12	10-14	

Mainly under long term contracts between producers and consumers. Some by-product output but mainly co-product in beach sands.

MARKETING ARRANGEMENTS

70% of world production of zircon sand from Australia. Metal production is in Japan, France and the US.

Index Numbers 1981 = 100

The solid line gives prices in money terms and the dotted line gives prices in 'real' 1981 terms

ZIRCONIUM
Zircon sand, 66–67% ZrO$_2$

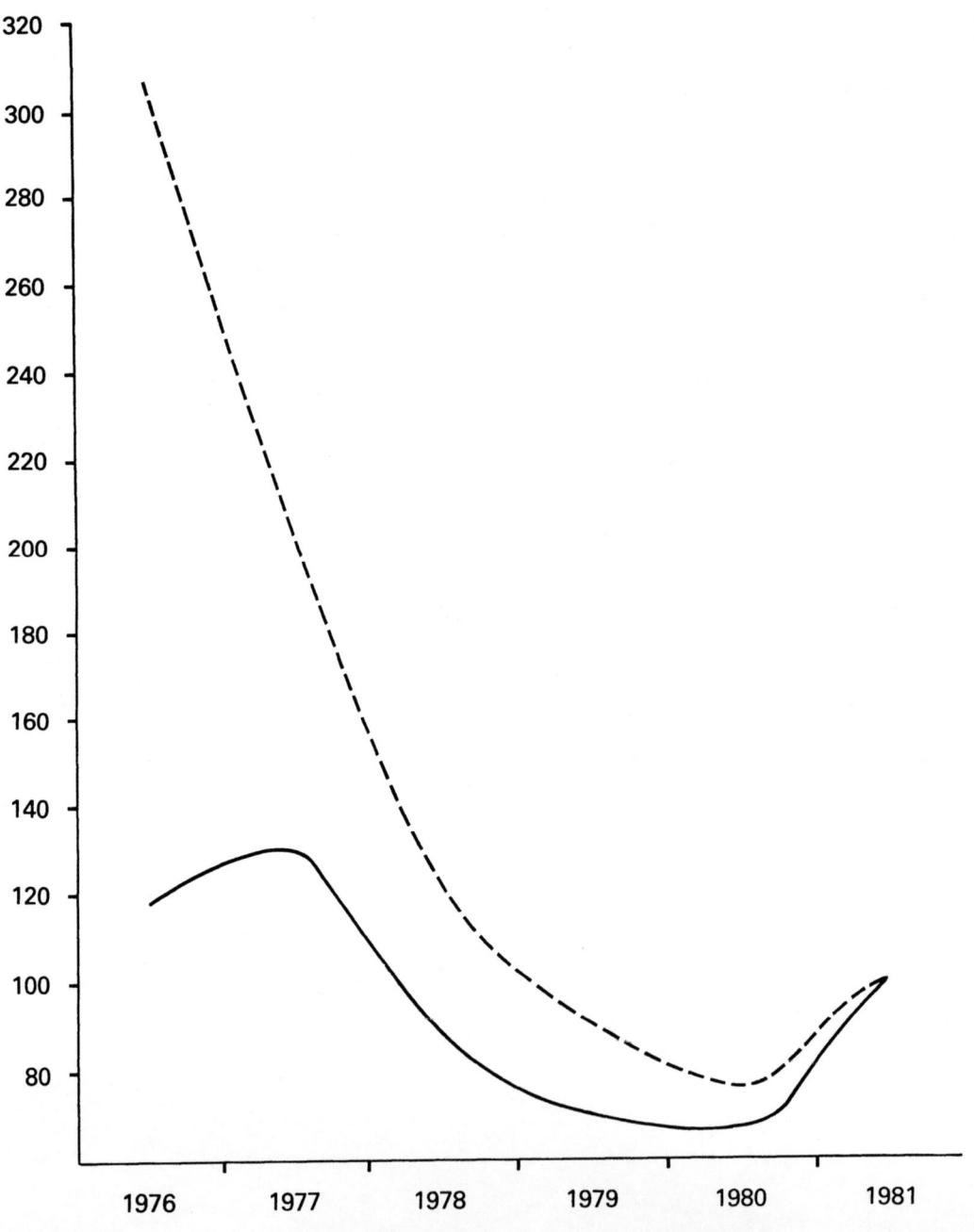

SUPPLY AND DEMAND BY MAIN MARKET AREA

	UK	EC (Ten)	Japan	USA
Production (1979/80 Averages) '000 tonnes				
Zircon concentrates	-	-	-	c.110
Zirconium oxide	n/a	n/a	3.48	9.68 (exc. metal producers)
Zirconium sponge and metal	n/a	n/a	0.36	n/a
Net Imports (1979/80 Averages) '000 tonnes				
Zircon concentrates	33.0	195.4	175.8	101.9
Zirconium oxide	0.15 (inc. germanium)	0.46 (inc. germanium)	n/a	0.29
Zirconium sponge and metal	0.2	1.23	0.12	0.74
Zirconium compounds	n/a	n/a	n/a	0.79
Source of Net Imports (%)				
Ores and concentrates				
Australia	76	81	88	88
S Africa	23	16	11	10
Other	1	3	1	2
Net Exports (1979/80 Averages) '000 tonnes				
Zircon concentrates	1.7	10.1	-	7.5
Zirconium oxide	1.04 (inc. germanium)	0.83 (inc. germanium)	-	1.76
Zirconium metal	..	0.85	-	0.73
Consumption (1979/80 Averages) '000 tonnes				
Zircon concentrates	30 to 35	c.190	c.175	141.5
Zirconium oxide	n/a	n/a	3.75	11.5
Zirconium metal	n/a	n/a	0.37	c.4 to 5

	UK	EC (Ten)	Japan	USA

Import Dependence (based on zircon concentrates)

| Imports as % of consumption | 100 | 100 | 100 | c.72 |
| Imports as % of consumption and net exports | 100 | 100 | 100 | 68 |

Share of World Consumption of concentrates %

| Total world (approx.) | 4 to 5 | c.25 | 20 to 23 | 19 |

Consumption Growth % p.a.

| All forms 1970s as concentrate | -1.7 (based on imports of conc.) | 4.2 (based on imports of conc.) | 8.4 (based on imports of conc.) | 0.2 |

****

SOURCES AND NOTES

This handbook has drawn from a very wide range of primary and secondary statistical sources that are almost too numerous to mention. Frequently, different sources may give markedly different estimates for what is ostensibly the same figure, even when full allowance is made for differing definitions. In such instances judgement has been used. Widely varying units of measurement are used in the originals, and this handbook has standardised, with one or two exceptions on metric units. Some errors may regrettably have crept into the conversions.

The main sources were as follows.

World Reserves

For nearly all minerals the data are taken from US Bureau of Mines' sources, and mainly from:-

> Mineral Facts and Problems. 1980 Edition. Bureau of Mines Bulletin 671
> Mineral Commodity Summaries 1981 and 1982
> Mineral Commodity Profiles (on specific minerals)
> Minerals Yearbook Volume 1. Metals and Minerals 1980

Production, Consumption and Trade

The same sources as for reserves, supplemented by:-

> World Mineral Statistics (various issues) published by the Institute of
>   Geological Sciences
> United Kingdom Mineral Statistics (successive issues) - Institute of
>   Geological Sciences
> Metal Statistics (successive issues) - Metallgesellschaft, particularly
>   for the main non-ferrous metals
> World Metal Statistics (successive issues) - World Bureau of Metal
> Statistics both for major non-ferrous metals and for some of the mines
> by-product metals
> Annuaire Minemet 1980 - Groupe Imetal
> Non-Ferrous Metal Data (successive issues) American Bureau of Metal
>   Statistics
> Metal Bulletin Handbook (successive issues) - Metal Bulletin
> United Nations Monthly Bulletin of Statistics
> Iron and Steel Statistics - (successive issues) - Eurostat
> Lead and Zinc Statistics (successive issues) - International Lead and
>   Zinc Study Group
> International Tin Council Monthly Statistical Bulletin (successive
>   issues)
> Ferro Alloy Statistics 1975-80 - Metal Bulletin
> Overseas Trade Statistics of the United Kingdom
> Foreign Trade Analytical Tables for 1979 and 1980 - (NIMEXE) - Eurostat
> Statistics of Japanese Imports and Exports for 1979 and 1980
> United States Trade Statistics for 1979 and 1980
> OECD Analytical Trade Statistics for 1979 and 1980 for Greece
> Yearbook of Mining, Non-Ferrous Metals and Products Statistics - M.I.M.
> Mining Year Handbook - M.I.T.I.

TEX Report - Ferro-Alloy Manual
TEX Report - Iron Ore Manual
The Rare Metal News
Yearbook of Fertilisers (Japanese)
Sumisho Non-Ferrous Metal News - Sumitomo Shoji K.K.
Metal News. The Japan News Industry Association
Aluminium, Copper, Lead and Zinc World Flow Tables - World Bureau of Metal Statistics
Uranium. Resources, Production and Demand (successive issues) - OECD Nuclear Energy Agency/IAEA

Prices

Metals Week Handbooks   1976-1980 + Metals Week   through 1981
Metal Bulletin Handbooks 1977-1981 + Metal Bulletin through 1981
Industrial Minerals
Engineering and Mining Journal
USBM Mineral Yearbooks and Mineral Commodity Summaries
IMF - International Financial Statistics
UN - Monthly Bulletin of Statistics
Sulphur - B.S.C.

Prices are yearly averages unless otherwise stated.

End Use Patterns

Mineral Facts and Problems. 1980 Edition. Bureau of Mines Bulletin 671
Mineral Commodity Summaries 1981 and 1982

Reserve/Production Ratios

The forecasts of cumulative demand underlying the 'dynamic' reserve ratios were taken mainly from Mineral Facts and Problems. In a few cases the US Bureau of Mines' forecasts were altered where they appeared unreasonable.